DDT AND THE AMERICAN CENTURY

THE LUTHER H. HODGES JR. AND
LUTHER H. HODGES SR. SERIES ON
BUSINESS, SOCIETY, AND THE STATE
WILLIAM H. BECKER, EDITOR

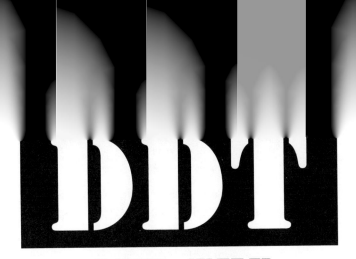

DDT

AND THE
AMERICAN
CENTURY

Global Health, Environmental
Politics, and the Pesticide
That Changed the World

DAVID KINKELA

The University of North Carolina Press
Chapel Hill

© 2011 The University of North Carolina Press
All rights reserved
Set in Whitman, Stencil, and Franklin Gothic
Manufactured in the United States of America

The paper in this book meets the guidelines for permanence and durability
of the Committee on Production Guidelines for Book Longevity
of the Council on Library Resources.

The University of North Carolina Press has been a member of
the Green Press Initiative since 2003.

Library of Congress Cataloging-in-Publication Data
Kinkela, David.
DDT and the American century : global health, environmental politics,
and the pesticide that changed the world / David Kinkela.
p. cm.
(The Luther H. Hodges Jr. and Luther H. Hodges Sr.
series on business, society, and the state)
Includes bibliographical references and index.
ISBN 978-0-8078-3509-8 (cloth : alk. paper)
1. DDT (Insecticide)—History—20th century. 2. Insect pests—
Control—History—20th century. 3. DDT (Insecticide)—
Environmental aspects.
I. Title.
SB952.D2K56 2011
632'.9517—dc23
2011015416

Portions of this work have appeared previously, in somewhat different
form, as "The Ecological Landscape of Jane Jacobs and Rachel Carson,"
American Quarterly 61, no. 4 (December 2009): 905–28, © The American
Studies Association; and "The Question of Success: Revisiting the
DDT Controversy from a Transnational Perspective, 1967–72,"
Ethics, Place & Environment 8, no. 2 (June 2005): 159–79.

15 14 13 12 11   5 4 3 2 1

To Cristina

Contents

Illustrations

Acknowledgments

I am indebted to a great number of people. Highly cognizant of my own inability to remember all those who have helped, I am privileged to acknowledge the many people whose contributions have been invaluable. And for those I have mistakenly neglected, my sincere apologies; your names may not be acknowledged, but your assistance remains etched in the pages of this book.

Throughout my research I have been helped by a number of terrific archivists. The professional staff at the Rockefeller Foundation Archive Center were extraordinarily helpful, particularly Erwin Levold, whose familiarity with the archive's expansive holdings made my time at the archive center extremely enjoyable and productive. At the National Archives in College Park, Maryland, an overwhelming place for many researchers, Mike Hussey and Joe Schwartz, among others, helped me navigate the voluminous records. William McNitt at the Gerald R. Ford Presidential Library was equally helpful. At the Lyndon Baines Johnson Presidential Library, Allen Fisher was a tremendous and enthusiastic guide, providing me with some incredibly valuable sources. Ben Rogers and Benna O. Vaugha at the W. R. Poage Legislative Library, a hidden jewel at Baylor University, were incredibly supportive and helpful. This book could not have been written without their collective efforts.

I would also like to thank the staffs at the Special Collections at the National Agricultural Library, Stony Brook University Special Collections, the New York Public Library and Science, Industry, and Business Library, Widener Library and Francis Loeb Library at Harvard University, Bobst Library at New York University, the Harry Ransom Center and Benson Latin American Collection at the University of Texas, and the interlibrary loan office at SUNY Fredonia.

This project has been supported by a number of grants, including a Rockefeller Archive Center grant-in-aid, the Gerald R. Ford Foundation Research Travel Grant, the Scholarly Incentive Research Grant and the Amy Elizabeth Everett Memorial Award at SUNY Fredonia, and a grant from the John and

J. D. Dowdy Memorial Congressional Research Endowed Fund at the Poage Legislative Library.

I have also been helped by two residential fellowships during which the bulk of this book was completed. The first was at the Institute for Historical Studies at the University of Texas. Under the estimable leadership of Julie Harwick, the institute was as intellectually stimulating and supportive a place as anyone could imagine. During my stay, Courtney Meador helped me navigate the administrative channels of the University of Texas and was a constant source of exuberance and joy. I would like to thank the other fellows, Jim Sweet, Ruben Flores, Nancy Appelbaum, and Ebru Turan, for making the experience professionally and personally rewarding. Also, thanks to the many people at the University of Texas for their encouragement, criticism, and wisdom, particularly Laurie Green, Frank Guridy, Anne Martinez, Tracy Matyisk, Mark Melzer, Robert Olwell, Seth Garfield, Susan Deans-Smith, Jorge Cañizares-Esguerra, Martha Newman, Antony G. Hopkins, Bruce J. Hunt, John Mckiernan-González, Alan Tully, Paul Rubinson, Jonathan Hunt, and Karl Hagstrom Miller. A special thanks to Mark Lawrence and Erika Bsumek, who have both been a source of unyielding support and friendship.

I also had the great pleasure of completing this book while serving as a fellow at the Charles Warren Center for Studies in American History at Harvard University. I would like to thank Sven Beckert, Erez Manela, and Nancy Cott for their guidance and interest in my project. Arthur Patton-Hock and Larissa Kennedy were extremely helpful throughout my stay in Cambridge, and I thank them both. Christopher Capozzola, Sandra Comstock, Dayo Gore, Sheyda Jahanbani, John Munro, Scott Nelson, and Sarah Phillips were incredible colleagues, making my stay at the Warren Center a truly wonderful and enriching experience. I learned so much from each of them; my debt can never be repaid.

I am grateful to a number of people who have been instrumental in my career. To Tom Bender, for his brilliance and unyielding support of my work. He has seen this project evolve in its various manifestations, yet throughout he has been forthright, critical, and encouraging. More than anyone, he has been steadfast in his belief in my work, and I cannot thank him enough.

Danny Walkowitz and Greg Grandin provided much-needed guidance as this project evolved and have been an incredible source of support since then. And Chris Otter, at an early stage, challenged me to write a better book. Thanks to you all.

While at New York University, I had the good fortune to work with and learn from some terrifically smart people, including Michael Lacombe, Kathleen M. Barry, Andrew Lee, Micki McElya, Stephen Mihm, Dan Prosterman, Marci Reaven, Ellen Noonan, Dan Bender, Kevin Murphy, Dayo Gore, Elizabeth Esch, Taja-Nia Henderson, Suzanna Reiss, Melina Pappademos, Felice Batlan, George Tomlinson, Aisha Finch, and Kim Gilmore. Neil Maher has been incredibly helpful throughout the process of writing this book, and I thank him for his constant encouragement. And to Noah Gelfand, who has learned more about DDT than he cares to admit, I offer thanks for his many contributions to this book.

My thanks to Tom Robertson, Paul Rossier, Sabine Clarke, Martin Melosi, William McAllister, Suzanne Moon, John Herron, and Jeff Sanders, who, in ways big and small, have all made contributions to this book.

I have benefited from the friendship and insights of Jimmy McWilliams, who was one of the great discoveries of my moving to Texas. One of my fondest memories of working on this project was talking about insects and history with him at the Dog and Duck.

Members of the department of history at SUNY Fredonia have been equally supportive. Particular thanks go to Ellen Litwicki, John Staples, Jacky Swansinger, and Mary Beth Sievens for their leadership and unwavering encouragement throughout. Emily Straus, Steve Fabian, and Zhao Ma, who read a number of draft chapters, provided much-needed feedback and advice. I value their friendship.

I would also like to thank Trisha Brandt Fox, Pietro Nivola, Frederick Toelle, and Alberto and Carlotta Guareschi for their time and interest in this project.

My editors, Mark Simpson-Vos and Ron Maner, have been wonderful to work with. Mark has helped me in countless ways, not simply by easing me through the publication process, but by being a wonderful sounding board as I presented him with one crazy idea after another. He has read and commented on a series of drafts, shepherding this book to its publication. I would also like to thank my copyeditor, Jeff Canaday, for making this book more readable; my thanks as well to the anonymous reviewers, who encouraged me to write a stronger book from start to end.

I would be remiss if I did not mention Susan Ferber at Oxford University Press and William Bishel at the University of Texas Press for their help throughout the publication process. They both have shared their thoughts on what this book might become and have made it better because of their counsel.

This book could not have been written without support from so many friends and family members. To Jeff Poulos and Brad Peloquin, thank you for your kind hospitality during my stay in Boston. To Pete and Sonia Lackey, Sue Choquette, and Jay and Julie Ritchie, thank you for providing me with warm shelter and good food. While in New York, I was put up on more than one occasion by Erin Courtney, Scott Adkins, and Homer Frizzell. To my great neighbors Ronald A. Romance and Tricia DeJoe, who, since my family and I arrived in Fredonia, have been more help than they can possibly know, my sincere gratitude. And to the many other friends who have refrained from asking, "How's the book coming?" thank you.

Lisa and Hans Gegenschatz have also been extraordinarily caring through-out the writing of this book. Having lived through much of the history this book explores, they have added a unique perspective that has been invaluable.

I am indebted to my parents, Dave and Sondra, and my sister Terry, not only for their unwavering inspiration and love, but also for making me aware of DDT at a young age. In the early 1970s in a cramped New York City apartment where space was minimal, we hung our "initialed" mugs underneath the lone kitchen cabinet. Rather than placing them alphabetically, D. D. S. T., we placed them in a particular order—S. D. D. T. In our family, the "S" stood for stop. It was a daily reminder that a larger world existed outside of our tiny apartment. And unbeknownst to me at the time, I would spend much of my adult life thinking about what it meant to "stop DDT."

And finally, to my sons Max and Milo and my wife, Cristina Gegenschatz, whose tireless efforts to keep things together made this book possible, I thank you. This project began when Max was just a baby, and it has been amazing to see him grow up as this book evolved. And since his arrival, Milo has been a constant source of joy. On more than one occasion, however, I had to close the office door in order to finish this book. I hope, in time, they come to understand the importance of their contribution. To my wife, Cristina, what more I can say? Thank you. I dedicate this book to you.

DDT AND THE AMERICAN CENTURY

DDT AND THE
AMERICAN CENTURY

In 1969, the World Health Organization (WHO) suspended one of the largest public health campaigns in world history. Established in 1955, the Global Malaria Eradication Programme was designed to rid the world of the parasitic disease at its source. Because malaria is transmitted by the bite of the female anopheles mosquito, the only mosquito genus capable of infecting humans with the disease, the public health campaign targeted the insect rather than the disease-causing plasmodium. In other words, malaria eradication required mosquito eradication.

Public health experts believed eradication would ultimately succeed in the historic struggle against malaria. And they had what many considered the perfect weapon, DDT. First synthesized in 1874, DDT had become the predominant method of insect control after the Second World War. DDT was inexpensive, easy to produce, effective, seemingly harmless to people, and it could be used in a variety of settings. For public health, the preferred method, what we now call indoor residual spraying, was a process in which health workers sprayed a fine mist of a DDT-petroleum mixture on the walls and ceilings of houses, public buildings, barns, and anywhere a mosquito might rest after taking its blood meal. Other techniques included spraying mosquito-breeding areas, ponds, swamps, or any body of standing water. To cover large areas, spray planes and helicopters were used, although this method was also deployed for agricultural, rather than for public health, purposes. Under the best circumstances, DDT residues remained effective up to six months after application. Between 1955 and 1969, millions of gallons of the chemical were used to annihilate mosquito populations, resulting in a sizeable decrease in malaria rates worldwide.

Despite these early successes, mosquitoes (and the malaria parasite) proved to be worthy adversaries, capable of not only withstanding the mammoth chemical assault but, over time, even prospering. Indeed, rather than surrender, anopheles mosquitoes developed resistance to the once powerful chemical and continued to reproduce in prodigious numbers and spread disease in human populations. So after fourteen years of waging war against malaria across much of the Global South, the WHO declared an end to the once promising public health project.

In the decades following the failure of the eradication campaign, malaria fighters struggled to develop a comprehensive program to combat the disease. Between 1970 and 2010, worldwide malaria rates skyrocketed. Those afflicted with malaria experience recurring bouts of debilitating fevers or die. Today, malaria infects nearly 30 to 40 million people annually, leading to over 1.5 million deaths a year. Children under the age of five are most susceptible to the disease; malaria is one of the leading causes of childhood deaths in Africa and other malaria-endemic regions.

Not to be deterred, in 1998 the United Nations Development Programme, the World Bank, and the WHO launched another comprehensive malaria project in conjunction with the United Nations Children's Fund called Roll Back Malaria (RBM). Rather than attack the anopheles mosquito, the new campaign was designed to protect human bodies from mosquitoes and the plasmodium using insecticide-treated bed nets, preventative antimalarial drug therapies, and early detection and treatment programs. DDT was universally rejected as part of the new project.

Unlike the earlier eradication programs, RBM projects focused predominately on sub-Saharan Africa, a region with some of the highest poverty and malaria rates in the world. Malaria, the president of the Togolese Republic declared, "heads the list of major endemic diseases that decimate the populations of Sub-Saharan Africa, causing desolation and hindering the development of the continent."[1] While malaria affects other areas of the world, the size and scope of the problem in Africa dwarfs those in other endemic regions. Recently, additional contributions from the Global Fund to Fight AIDS, Tuberculosis, and Malaria; President George W. Bush's Malaria Initiative; and the Bill and Melinda Gates Foundation have pushed total commitments to $1.7 billion in 2009, but these contributions are still well short of the $6 billion estimated to be necessary for the global implementation of malaria-control programs. Despite improved efforts at disease prevention, mosquitoes continue to infect people with disease-causing plasmodium with alarming frequency.[2]

IN 2012, ENVIRONMENTALISTS will celebrate the fiftieth anniversary of the publication of Rachel Carson's *Silent Spring*, a book that, because of its social and political impact, has been considered the *Uncle Tom's Cabin* of its day. Transformative as it was informative, *Silent Spring* provided the nascent environmental movement of the 1960s with a set of ideas that resonated widely. Perhaps most important was the concept of ecology, a science predicated on the study of the interaction and exchange between animals, plants, and environments, what we now call an ecosystem. Carson's rich prose and analytical sophistication unravel nature's complexity into easy-to-understand concepts. And her readers loved it. *Silent Spring* quickly jumped to the top of the best-seller list, becoming one of the most important books of the twentieth century, even if it was about a particularly banal topic, pesticides.

The villain of Carson's story was DDT, a seemingly innocuous chemical, but one, Carson warned, with hidden dangers. Indeed, to many Americans, DDT posed little risk. Its widespread use since the Second World War revealed little evidence that the chemical was harmful to humans. Except in a few cases of wanton misuse, DDT was not acutely poisonous, nor did it seem to have any long-term risks. The same could not be said for other pesticides that emerged on the market, many of which were arsenic based or derived from German nerve gas experiments; a mistake in handling these chemicals could be deadly. The lack of immediate harm masked any long-term dangers of DDT. Even today, proponents of the chemical refer to DDT as an "excellent powder."[3]

A number of Americans, including Rachel Carson, were suspicious of this calculus. It became apparent that the massive spray programs produced unforeseen problems. In areas sprayed with DDT, dead birds and fish began appearing more frequently, sometimes in startlingly high numbers. Even before DDT was made available to the public, on August 9, 1945, the *New York Times* reported that experimental tests for mosquito control in New Jersey "showed that fish were killed in quantities."[4] By 1950, ecologists and wildlife biologists confirmed that the main culprit was the chemical many believed to be risk free. Over the following decade, additional studies exposed the ecological consequences of DDT, creating a consensus among a small group of biological scientists about the inherent dangers of the chemical. Carson's impetus for writing *Silent Spring* was to bring these scientific studies to a wider audience, an ambition she notably achieved.

Publication of *Silent Spring* led to a flurry of diverse and derisive responses. Many Americans accepted Carson's work, believing that she unveiled the invisible dangers of the postwar world. Others rejected her conclusions outright, suggesting that a woman could not fully understand the complexity of modern science and do so rationally or without emotion. Still other critics believed Carson undermined the foundation of American postwar supremacy with her audacity to question the scientific and technological innovations that sustained the most robust economic and military power in the world. Upon its publication, the ruckus over *Silent Spring* was anything but silent.

By 1972, the conflict over DDT had come to a head. Not only had the suspended WHO project revealed the limits of DDT, but the once nascent environmental movement had matured, creating a formidable political force capable of shifting the balance of power in Washington. Shortly thereafter and after a lengthy legal battle in which advocates and critics attempted to define the meaning of DDT in a changing world, the Environmental Protection Agency (EPA) banned the domestic use of the pesticide. The EPA did not, however, ban the export of DDT or its use for public health programs overseas. Other nations followed the United States in banning the chemical, and by the end of the century only a handful of nations produced the chemical for a shrinking, yet largely unregulated global market.

Fearing the ecological fallout from unregulated DDT use, in 2001, the United Nations drafted the Stockholm Convention on Persistent Organic Pollutants, also known as the POPs Convention, an international agreement to limit the use of DDT and prohibit other "persistent pesticides." According to the convention, a ban was needed because these pesticides "possess toxic properties, resist degradation, bioaccumulate and are transported, through air, water and migratory species, across international boundaries and deposited far from their place of release, where they accumulate in terrestrial and aquatic ecosystems."[5] The POPs Convention represented a profound shift in controlling the flow and use of persistent pesticides on a global scale. As the earlier EPA ruling had, the agreement made exceptions for DDT used in public health programs, even though funding for these programs had dried up or been diverted to RBM programs. By 2010, 152 of 170 U.N. member nations ratified the POPs Convention. The United States, along with Russia, Italy, Saudi Arabia, and Ireland, among others, has yet to endorse the treaty.

Nevertheless, during the 1990s, a few isolated voices began to sing the praises of DDT. Michael Fumento and the conservative columnist Michelle Malkin, then fellows at the Competitive Enterprise Institute, condemned Carson for her "folly" in exposing the "dangers" of DDT in relation to malaria eradication projects. Writing about malaria control programs in Ceylon (now Sri Lanka), Fumento and Malkin scoffed, "But after spraying was stopped in 1964, as a direct result of Carson's book, malaria cases quickly shot back up."[6] In reality, suspension of that program had more to do with the success of the eradication campaign than with anything Carson wrote, and the program's failure resulted from DDT-resistant mosquitoes rather than some radical environmental agenda.[7] Nevertheless, a decade later, these lone voices erupted into a chorus, as more and more people began to question reductions in DDT use. Collectively and unapologetically they rejected Roll Back Malaria in favor of a renewed commitment to DDT spraying. Over the past decade, these voices have grown louder and more vociferous.

In a Senate hearing in 2005, Roger Bate, codirector of Africa Fighting Malaria and fellow of the American Enterprise Institute, told the Subcommittee on Federal Financial Management, Government Information, and International Security, "We appear to be losing the war on malaria. Far from rolling the disease back it appears to be increasing."[8] Paul Driessen, a senior fellow at the Atlas Economic Research Center and author of *Eco-Imperialism: Green Power, Black Death* (2006), vilified the environmental movement, claiming, "Once again, the environmentalists are willing to ban DDT because they are willing to sacrifice real human lives to theoretically save the lives of birds." J. Gordon Edwards, an entomologist and longtime supporter of DDT, and Steve Malloy, cofounder of the Free Enterprise Action Fund (now the Congressional Effect Fund), have claimed that efforts to prohibit DDT are based on "junk science."[9]

The Competitive Enterprise Institute operates a website called "Rachel Was Wrong." The site claims, "Millions of people around the world suffer the painful and often deadly effects of malaria because one person sounded a false alarm. That person is Rachel Carson."[10] Even the Congress of Racial Equality, historically one of the nation's most influential civil rights organizations (although politically at odds with its earlier incarnation), has condemned the DDT ban.

The condemnation continued. In his novel *State of Fear*, the late Michael Crichton asserted that the ban on DDT "killed more people than Hitler." In 2007, Oklahoma senator Tom Coburn blocked legislation commemorating Rachel Carson, claiming "Carson's condemnation of the pesticide DDT" has led to developing nations abandoning "the best weapon . . . available in the battle against malaria." And more recently the documentary film *Three Billion and Counting* took its title from sources making claims about the number of people infected with malaria since "the greatest ecological genocide in the known history of man . . . the 1972 ban of an extraordinary life-protecting chemical, DDT." As a group, these critics have used the case of DDT to condemn federal and international regulation, despite provisions making DDT available for public health measures.[11] The vociferousness, however, remains alarmingly exaggerated.

Underlying this contemporary debate is a complex history rooted in ideas about human health, science and technology, modernity, and state regulation. It is a history shaped by geography and nature and influenced by the presence of disease, famine, and insects in certain parts of the world and their relative absence in other regions. It is also a history about the meaning of U.S. power and global influence since the Second World War. More than just a chemical pesticide, DDT bears a complicated and contested history that has been reduced to a rather simplistic reading of the past. This book revisits the history of DDT, adding context and color to a picture largely devoid of nuance and defined primarily in black and white.

At its root, the contemporary debate involves contrasting ideas about the role of technology and ecological knowledge. Blaming environmentalists for the prohibition of DDT simply masks a much more contentious history regarding the use and misuse of technology, the scientific and ideological understanding of ecology, the impact and legacy of the Cold War, the legal challenge of environmental protection, and America's changing standing in the world.

This book traces the flow and use of DDT around the world while placing the United States, which more than any other nation encouraged the global use of the pesticide, at the center of the story. During the postwar period, American public health and agricultural development experts embarked on a series of development projects designed to solve the problems of disease and famine. Integral to what the famed magazine publisher Henry Luce called "the American Century," these development projects embraced the possibility of eradicating the problems of the past—disease, famine, and poverty— through the strategic deployment of modern technologies, including DDT.

Luce's influential 1941 essay, "The American Century," asserted that the United States should take its rightful place as "the Good Samaritan of the entire world." He believed in the transformative nature of American ideas, technology, and power. "As America enters dynamically upon the world scene," Luce wrote, "we need most of all to seek and to bring forth a vision of America as a world power which is authentically American." As part of this renewed faith in American exceptionalism, entomologists, public health officials, and agricultural development experts promoted DDT as an "authentically American" technology.[12]

DDT had its own history to write. Not only was it an effective pesticide, but it was mobile, persistent, and moved easily across borders and through the food chain. The hazards of DDT, therefore, were not bound by any one nation, but were, in fact, global. These environmental realities challenged prevailing notions of postwar development, highlighting the ecological interconnections between people, nations, and nature.

DDT became a symbol of two countervailing forces shaping American perceptions about the world and the place of the United States in it. Perhaps the most dominant force was an unyielding belief in technology as a marker of modernity. In this regard, DDT was just one of a multitude of postwar technologies that characterized the modern world.[13] During the postwar period, Americans also developed an awareness of the interconnections between the human and nonhuman world. By the 1960s, ecology emerged as a profoundly transformative idea that not only changed how humans interacted with the natural world, but also redefined what it meant to be modern. In his classic treatment of the evolution of postwar urbanization, the philosopher and urban scholar Marshall Berman called this ecological turn "antimodern modernism."[14] This book places these two concepts—technological modernity and ecological modernity—in dialogue with each other, tracing how these ideas shaped, transformed, and defined DDT and the American Century.[15]

GIVEN THE CHEMICAL'S EUROPEAN ORIGIN, it is somewhat surprising that DDT became such an important American technology. Developed in 1939 by Paul Müller, who worked for the Swiss chemical company Geigy A.G., DDT quickly came to prominence during the Second World War. Even though both sides in the conflict procured samples of the compound, only the United States established institutional mechanisms to determine how and whether to use DDT extensively during the war.

For American scientists, DDT posed a unique problem. Not only was little known about the chemical, but prior to the war DDT had been used primarily as an agricultural pesticide, so its effectiveness against disease-bearing insects remained unclear. Moreover, researchers had to develop viable methods to deliver DDT under wartime exigencies, a problem that proved challenging yet surprisingly straightforward. American entomologist Edward Knipling later recalled, "DDT just changed things. I mean it was a revolution in control of insects affecting man and animals."[16]

Indeed, the war changed the meaning of DDT. It was no longer an unknown chemical; it was a miracle compound that saved the lives of millions of people. It was also a uniquely American pesticide, produced by some of the largest chemical manufacturers in the United States for an eager domestic market. On September 9, 1945, Gimbels, the New York department store, ran a full-page ad in the *New York Times* claiming the store had the first shipment of DDT, "Released Yesterday! On Sale Tomorrow! Gimbels Works Fast! . . . You know what wonders DDT worked for the armed forces," the ad reads, enticing housewives to buy the new "Aer-A-Sol bomb." Soon, many Americans bought, sprayed, dusted, fogged, and applied DDT as the first defense against insect pests.

DDT was also part of a larger scientific revolution that would affect the postwar world. During the war, Vannevar Bush's Office of Scientific Research and Development directed some of the most significant scientific research of the twentieth century, including the development of sonar, radar, penicillin, and the atomic bomb. By the war's end, Bush continued to champion science as a state-security issue. In his influential report from 1945, "Science: The Endless Frontier," Bush wrote, "Since health, well-being, and security are proper concerns of Government, scientific progress is, and must be, of vital interest to Government."[17]

The Cold War made Bush's pronouncements seem prophetic. Indeed, the ideological and political struggle that enveloped nearly every part of the world converged around questions of national security and global health. Despite periods of violence, for many people around the world, Cold War science improved health and agricultural production on a global scale, raising living standards for some of the most impoverished people in the world. Not only was DDT critical to these development projects, but it was also crucial to the geopolitical aspirations of American cold warriors and advocates of the American Century.

The American Century melded different interests in a common purpose. Business leaders, philanthropic organizations, media outlets, U.S. foreign policy officials, and a diverse network of scientists forged new alliances to promote American interests abroad. The effort was not collaborative in the sense that it was unified, but it was multifaceted and diverse, based on a belief in American superiority. It drew its strength from various state and nonstate actors promoting "authentically American" ideas and goods to people around the world, goods that became particularly valuable as Cold War lines hardened.

Encouraged by a commitment to remake the world in an American image, business leaders, public health officials, agricultural experts, and government officials traveled far and wide to spread the gospel of American technology. While the first large-scale demonstration of DDT's effectiveness took place in Europe, American interests would shift to the "Third World," where DDT was a critically important technology in remaking the expansive and incredibly diverse region stretching from Latin America to Southeast Asia. In an effort to break the cycle of poverty, malnutrition, and disease in much of the Third World, development experts put their faith in science, technology, and American capitalism to reverse historically "backward" economies. To modernize the unmodern, Americans vigorously promoted technology as the key to unlocking the past. In efforts ranging from the Marshall Plan to the U.S. Agency for International Development, the United States sponsored a series of foreign aid programs to stimulate economic development and advance American interests abroad.

International aid organizations also played a central role in promoting DDT and U.S. interests abroad. Undoubtedly the most influential was the Rockefeller Foundation, an organization working at the vanguard of scientific knowledge and strongly promoting its expertise, money, and technology as mechanisms for social and political change.[18] The foundation's use and promotion of technology was based on the central question of modernity. What it meant to be modern, however, varied significantly as different nations held differing and, at times, conflicting ideas about technology and development. For the foundation, these issues were rather straightforward because science and technology lay at the heart of its institutional mission. And not surprisingly, DDT was critical to the foundation's modernizing project.

To examine this history, this book crosses many boundaries, from the rugged landscape of Sardinia to experimental stations in Chapingo, Mexico; from the political capital of the United States, Washington, D.C., to the agricultural fields in India; and from the backyards of American suburbanites to the malarial regions in Africa, as a way to tell a complex story. And while this story traverses multiple terrains, it is essentially a narrative that explores a global history from an American perspective.

This book examines the role the United States had in sponsoring the rise of chemical pesticides overseas and seeks to understand the various meanings of U.S. environmental policy in a global context. It explores how scientists, philanthropic foundations, corporations, national governments, and transnational institutions assessed and adjudicated the balance of risks and benefits of DDT within and beyond U.S. borders. And it considers how American environmentalists understood the global implications of DDT, particularly the ways in which DDT related to public health and agricultural development. This book, therefore, does not tell a global history of DDT per se, but places the American history of DDT in an international context.[19]

By investigating the flow of ideas, money, people, and, in this case, pesticides across borders, we can better comprehend the complexity of DDT's history and its connection to the history of the American Century. Recently, historians of the modern period have transcended borders, suggesting that the boundaries of the nation-state hide more than they reveal. For their part, environmental historians have been at the forefront of the historiographic shift, contending that history can (and should) be written "without borders."[20] In posing such a research agenda we must also be cautious because, as this book attests, borders do matter, politically, economically, and environmentally.[21] Even as historians take a transnational turn in an effort to develop a clearer image of the past, the nation-state, especially in the postwar period, played a significant role in how people understood their world.[22] This book, therefore, does not give equal treatment to regions outside the United States. Rather it explores how American scientists, political activists, philanthropic organizations, and government officials understood the world beyond U.S. borders and the place of DDT in it.

This approach refocuses the historical lens on the complicated and contentious history of DDT. By looking within and beyond the political borders of the United States, this book not only sheds new light on the international dimensions of DDT's contested history, but opens new terrain to reexamine the formation of U.S. environmental policy. In essence, this approach shifts the focus away from a simple narrative that shapes contemporary discourses about DDT, environmental policymaking, and the role of technology in the world to call attention to how people across borders understood the risks and benefits of DDT and based their understandings on different ideas about technology, ecology, and the role of the state. "It makes no sense to conceive of this interaction as a clash between modern and traditional sectors, or between the principles of movement and stasis," historians Charles Bright and Michael Geyer wrote in their influential essay, "Where in the World Is America?" Instead, they submit, "global practices encode local contexts, but they are also remade by them."[23] Thus, as American experts deployed DDT around the world, what happened beyond U.S. borders had a profound impact within them. The domestic use of DDT, similarly, influenced its deployment overseas. Rather than being separate and distinct—one use foreign, one domestic; one use technological, one ecological—domestic and international uses of DDT were intertwined.

Television, radio, newspapers, and magazines carried news of American efforts to reshape the Third World. Depictions of ill-health, famine, overpopulation, and the ever-present threat of communism shaped how Americans came to understand the impact of DDT throughout world. Historian Melani McAlister suggested that during the Cold War Americans "encountered" foreign people and places "through a medium of culture."[24] This was certainly the case with DDT, as press reports spoke glowingly about the transformative power of the chemical. Cold War tensions only heightened Americans' expectations about DDT, and the American press was more than willing to confirm their desires. Technological enthusiasts, cold warriors, and the American press created a powerful set of ideas that gave meaning to DDT for American audiences. As early as 1945, ecologists challenged these press accounts. But, as we shall see, these press reports would be difficult to overcome because of the contentious political climate in which they were generated.

The domestic and international use of DDT sparked a series of questions about the limits of liberal governance within the United States. How or whether governments were to regulate the use of chemical pesticides was a significant question in its own right, but the broader political context in which decisions related to that question were made enables us to rethink the emergence and impact of environmental governance. For example, environmentalists have celebrated the prohibition of DDT within the United States as the highpoint of liberal environmental governance. Indeed, the regulatory framework created during the 1960s and 1970s indicated the nation's willingness to deal with a myriad of environmental problems, including the ecological hazards of DDT. Unlike more localized hazards of pollution, resource destruction, smog, and water quality, the global impact of DDT raised a number of complex questions about the limits and consequences of federal regulation.

Furthermore, because the use of DDT was so enmeshed with the politics of the Cold War, the debate over pesticide regulation took on greater significance to all who participated. The debate did not merely involve local or national questions, but questions about the role of the United States in the world. DDT regulation took on a set of questions that challenged Americans' perception of the United States and the world. This book, therefore, places DDT's history in a broader framework that both recognizes political borders and transcends them.

DDT REMAINS A CONTESTED TECHNOLOGY, and it will be as long as insects and humans coexist. Rather than prescribe a solution to these contentious debates, this book unpacks the historical contexts in which the debates took place. It widens the lens of historical analysis, looking at the interconnection between national and international history, an interconnection that has shaped and continues to shape the story of DDT.

1

AN ISLAND IN A SEA
OF DISEASE
DDT Enters a Global War

*It should not be inferred from the results of these surveys that DDT is
not all that its champions claim it to be, an almost perfect insecticide.*
—Fred Soper, W. A. Davis, F. S. Markham, and L. A. Riehl,
The American Journal of Hygiene, 1947

The Japanese bombing of Pearl Harbor propelled the United States into a
global war. It also set in motion a series of transformative events that redefined the nation. Millions volunteered for military service, including thousands of Filipino Americans, Japanese Americans, and African Americans
who navigated the complexities of military service, racial discrimination,
and citizenship. Millions of other Americans, including women, supported
the war effort through their labor. Domestically, the Arsenal of Democracy
afforded women and minorities real, albeit temporary, employment opportu-
nities within the industrial sector. Massive public investment in defense also
reversed the nation's economic fortune, which had been mired in a decade-
long depression, setting the stage for a remarkable period of prosperity after
the war. And, under the leadership of Vannevar Bush, President Roosevelt's
scientific adviser, the United States strengthened its commitment to science,
forging new networks of knowledge between federal agencies, business, and
universities that would eventually bolster the nation's Cold War strategy.

What was surprising about the enormous deployment of soldiers and civilians at the onset of war, and what has been largely overlooked, was the health
of those deployed. By 1941, the current generation of Americans was arguably
the healthiest population in the nation's history. Despite the hardships of

the Great Depression, average life expectancy increased from around forty-nine years in 1900 to nearly sixty-four at the start of the war.[1] Advances in medicine, disease prevention, and, more critically, public health ushered in a new era of health in the United States. Consumer and workplace protections ensured that foods and factories were safer. Rural and urban development projects under the New Deal contributed to the reduction of communicable diseases. The Farm Security Administration, for one, devoted enormous resources to public health. Other alphabet agencies were similarly committed to health.[2] While discriminatory practices continued to limit access to quality health care, by 1941 Americans were increasingly, if not overwhelmingly, a healthy group of people.

Even though polio remained an unexplained horror and cancer rates increased throughout the century, a host of other human ailments receded into the past. Some of the largest killers of the nineteenth century—tuberculosis, cholera, and diphtheria—had either been expunged or were in steep decline.[3] Insect-borne diseases like typhus, yellow fever, and malaria had been reduced to historic lows. According to a survey conducted by the Office of Malaria Control in War Areas, a wartime predecessor to the Centers of Disease Control and Prevention, "The incidence of malaria in the United States at the beginning of the war emergency [was] the lowest in history."[4]

Thus, as Americans readied for war, they did so with firm resolve *and* sound body. The marked improvement in health prior to the bombing of Pearl Harbor underscored the nation's ability to wage war on two fronts.

Looking beyond the nation's borders, however, a different story emerged. Military deployment sent U.S. servicemen and -women to some of the most dangerous environments in the world. Countless diseases threatened American forces throughout the Mediterranean and South Pacific, including filariasis, typhus, dengue fever, and malaria, an affliction that concerned Douglas MacArthur, the supreme allied commander in the Pacific. "This will be a long war," MacArthur proclaimed, "if for every division I have facing the enemy I must count on a second division in hospital and a third division convalescing from this debilitating disease."[5]

MacArthur's concerns were certainly well founded. Not only did U.S. servicemen and -women endure difficult circumstances against dangerous enemies, but sickness also undermined their ability to fight effectively. MacArthur fully understood, as did many top U.S. commanders, that disease was as dangerous, if not more so, than a well-armed adversary. History bore this out.

"Typhus," the noted American bacteriologist Hans Zinsser wrote in his influential book *Rats, Lice, and History* (1934), "with its brothers and sisters,

—plague, cholera, typhoid, dysentery,—has decided more campaigns than Caesar, Hannibal, Napoleon, and all the inspector generals of history."[6] In 1780, nearly half of General Cornwallis's British troops succumbed to malaria, giving aid to the outnumbered American soldiers in the War of Independence. Shortly thereafter, yellow fever killed 22,000 of the 25,000 French troops during the Haitian revolution. Typhus has been credited with helping Russia defeat Napoleon in 1812–13. In the First World War, British forces in Salonika lost 162,000 men to malaria, while there were only 28,000 battle casualties. And the global impact of the Spanish flu, which killed nearly 18 million people during the First World War, confirmed in the minds of many public health and military officials the deadly consequences of disease.[7]

The Second World War created its own set of challenges. Even under the best of circumstances, the struggle with disease proved quite vexing. Because the war was fought in deserts, tropical islands, temperate forests, swamps, and cities, health officials faced a variety of diseases that in their own right would be difficult to cure. The destructive impact of war made fighting these diseases a titanic struggle. Insects thrived in wartime environments. Breeding areas for mosquitoes multiplied, increasing the likelihood of epidemics of malaria, filariasis, and dengue fever for civilian and military populations. Those forced to live among the rubble of destroyed cities and towns faced enormous health challenges, including the threat of typhus, dysentery, and plague. It was not just a war; it was, according to Surgeon General Thomas Parran, "a public health war."[8]

Crossing the Atlantic or Pacific, therefore, meant more than just heading to the front. It meant entering environments where the presence of insects and disease threatened a soldier's ability to fight effectively. By the start of the war, the chasm between health in the United States and the unhealthy environments millions of soldiers would soon enter widened. Looking back at the war effort, Robert Moore, a professor of pathology at Washington University, remarked that the United States was "an island in a sea of disease."[9]

Building on the promising public health work of previous decades, the global war created a powerful moment in which entomological research, new methods of public health, and the development of new chemical pesticides coalesced around a common theme: that of protecting soldiers from insect-borne diseases around the world. Arguably, DDT was one of the most important scientific innovations of the war. Its significance rested not only on its capacity to kill insects—other chemicals were effective insecticides as well—but on its ability to reduce a series of complex biological problems into a single solution. Spraying DDT simplified public health strategies, making it the dominant method of insect control by the war's end. This shift was neither immediate nor universal. Instead, it reflected the work of a vast scientific network determined to solve a range of health-related problems beyond U.S. borders.

DDT WAS AN UNLIKELY HERO OF THE WAR. In 1939, Paul Müller, a relatively unknown scientist working for the Swiss chemical company Geigy A.G., made a remarkable discovery. Charged with finding an agricultural chemical to control the Colorado potato beetle, an invasive species that threatened the important Swiss crop, Müller uncovered a chemical formula developed nearly seventy years earlier, in 1874. At that time, Othmar Zeidler, a graduate student at the University of Strasbourg, synthesized a crystalline compound, combining chloral hydrate with chlorobenzene in the presence of sulfuric acid. Zeidler's innovation was dichloro-diphenyl-trichloroethane, more commonly known as DDT. Because Zeidler's synthesis was only an experiment to fulfill his thesis requirement, he did not recognize the insecticidal properties of the compound. Consequently, his work was shelved.

Years later, when Müller rediscovered the chemical formula, he was immediately amazed. Initial tests demonstrated DDT's efficacy against Müller's test subjects, houseflies. What amazed Müller more than DDT's immediate effectiveness was its persistency. Day after day, Müller returned to the laboratory to find dead flies in the experimental container that had been treated with DDT only once. Encouraged by these findings, Geigy immediately deployed DDT on infested potato fields. Almost magically, the Colorado potato beetle infestation ended within a year. By 1940, DDT proved to be an effective agricultural pesticide with seemingly limited or no toxicological impacts on humans.

But Müller's discovery generated little international publicity. Europe was engulfed in a continental war, and the United States remained an interested observer. But as the United States entered the war, which expanded across continents and into the tropics, where the threat of insect-borne diseases increased, military health officials on all sides of the conflict demanded new methods to control disease, and DDT was positioned to play an important role in the war effort.

At the start of the war, pyrethrum, a botanical compound derived from chrysanthemum petals, seemed to offer much promise. A popular and effective insecticide, pyrethrum had been used to control body lice during the Napoleonic wars and remained a widely used form of pest control in the United States. American chemical producers proclaimed that pyrethrum offered soldiers "Freedom from Pestilence." Enlisting "Jonny Pyrethrum," another U.S. manufacturer noted, demonstrated that the chemical industry was "doing a big job for Uncle Sam in controlling the pest that affects the health of our men on every war front—yes pyrethrum is on malaria patrol."[10]

Such optimism would be short lived. By 1939, the U.S. imported 13.5 million pounds of pyrethrum from two major producer-nations, Kenya and Japan, with Japan producing over 90 percent of the chrysanthemum blossoms.[11] The war brought an abrupt end to the Japanese market and, as the war progressed, Kenyan production could not keep up with demand. As a result, U.S. military officials looked to military and civilian scientists to develop an alternative insecticide.[12] Brigadier General James S. Simmons, army chief of preventive medicine, informed a group of civilian researchers that the most important thing they could do for military medicine was to find "a substitute for pyrethrum."[13]

Finding the substitute for pyrethrum would fall under the jurisdiction of the Bureau of Entomology and Plant Quarantine (BEPQ). Part of the U.S. Department of Agriculture (USDA), the BEPQ came into existence in 1872 (and was then the division of entomology) to coordinate control of invasive boll weevil and gypsy moth populations that threatened crops and native flora. Later, under the leadership of Leland Ossian Howard (bureau chief, 1894–1927), whose book, *Mosquitoes: How They Live; How They Carry Disease; How They Are Classified; How They May Be Destroyed* (1901), became a classic text in the field of entomology, the bureau moved from supporting biological and cultural insect control methods to recommending chemicals, including Bordeaux mixture, lead arsenate, Paris green, sulfur, and kerosene. At the start of the war, the bureau remained steadfast in its commitment to chemical pesticides.

Edward Fred Knipling was called upon to direct the laboratory of the Bureau of Entomology and Plant Quarantine in Orlando, Florida. A native of the coastal town of Port Lavaca, Knipling completed his master's degree in entomology from Texas A&M in 1932, and he spent much of the early 1930s on the move, working on a number of USDA projects in Iowa, Illinois, and Georgia. In 1937, Knipling returned to Texas and embarked on a series of research projects that focused on the biological control of screwworms, a parasitic fly that threatened livestock populations in Texas and along the Mexican borderlands region. In 1940 Knipling moved to Portland, Oregon, after being named the new director of the Division of Insects Affecting Man and Animals. Knipling was charged with investigating mosquito populations of the Pacific Northwest, a region and insect species that were quite foreign to him. His stay in Portland, however, would be short lived as war interrupted the best-laid plans of most Americans.[14]

Knipling's laboratory would become the epicenter of a massive cross-border research effort to determine how to protect people and crop plants from the dangers of insects. Yet the laboratory was not alone in its effort to improve human health. No stranger to the ravages of disease, President Roosevelt increased federal funding for health-related programs. Jonas Salk, for example, who would go on to be the celebrated inventor of the polio vaccine, spent the war years working on flu vaccines. The Public Health Service's Office of Malaria Control in War Areas, directed by Dr. Louis Laval Williams, developed preventative strategies to reduce the hazards of malaria at home and abroad. Roosevelt also funded the nascent Committee on Medical Research, which would spearhead biomedical research on antimalarial drugs. Additionally, in December 1942, Roosevelt established by executive order the United States of America Typhus Commission. The commission was designed for the "purpose of protecting the members of the armed forces from typhus fever and preventing its introduction into the United States." Operating under the command of the secretary of war, the commission was given free rein to construct public health programs to protect military personnel from typhus "within and without the United States."[15]

The scientific and institutional linkages developed for disease prevention and control during the war were not limited to the United States; they were in fact global. Rather than being an isolated research station, the Orlando laboratory was part of an international network of scientists focusing on limiting the impact of insect-borne infections during the war. Research conducted by military and civilian scientists in Mexico, Brazil, Trinidad, Australia, Algeria, Egypt, and Italy developed new understandings of disease within different environments. This expansive scientific undertaking would ultimately reshape the postwar period. Yet, during the war, U.S. science remained steadfast in its commitment to vanquish foreign enemies, human and insect alike.

Knipling and his fellow scientists at the Orlando laboratory faced the sea of disease head-on and were determined to navigate and conquer the treacherous world that lay beyond U.S. borders. The enormous promise of DDT provided a means to envision this victory, but much was unknown about the chemical few had ever heard of. The question was whether DDT could fill the vacuum left by the undersupply of pyrethrum.

Despite the limits of initial wartime markets, the Swiss manufacturer Geigy attempted to capitalize on Müller's discovery. In 1940, Geigy began marketing two DDT products; one was a louse powder called Neocid, and the other a spray insecticide sold under the name Gesarol. Used primarily as agricultural agents within Switzerland, Neocid and Gesarol provided remarkable protection against a number of insect pests. Building on the success of early experiments, Geigy provided samples to Allied and Axis powers, and in so doing set off a series of experiments linking DDT to diverse ecologies, insects, and human bodies. It was a turning point for public health.

In 1942, Geigy sent 100 pounds of Gesarol to Victor Froelicher, the head of the company's New York subsidiary. Froelicher translated Müller's findings and began conducting further tests of the compound. He also sent a small sample to Knipling's Orlando laboratory.[16] Tests conducted by Knipling's research team on DDT paralleled Müller's findings. According to Knipling, DDT was "immediately recognized as the most toxic larvicide known, and intensive laboratory and field studies soon established that it was not only highly effective but very versatile in its uses."[17] Moreover, the pesticide seemed harmless to humans.[18]

Geigy also sent samples to German scientists, who undertook a series of experiments that both paralleled and departed from Knipling's research. Receiving their first DDT shipment in 1940, German agricultural experts thought they had found a magic bullet to protect crops from insect infestation. Given the importance of crop production during the war, many German agriculturalists believed that the Allied forces would use the potato beetle as a form of biological warfare to undermine the food production capacity of the nation. Conversely, German public health officials were less certain about the curative power of DDT. At a time when the healthy, vigorous German body seemed the ultimate manifestation of German nationalism, Nazi doctors questioned the risks associated with DDT. Theodor Morell, Adolf Hitler's personal physician, prevented the distribution of the pesticide until 1943, alleging that DDT was both dangerous and useless and had the potential to harm German bodies, thus compromising the health of the nation as well. So even as Nazi soldiers suffered from an outbreak of typhus during the battle of Stalingrad, German health officials rejected the potential benefits of DDT, even though initial tests revealed the pesticide's effectiveness.[19] Eventually, the German army began spraying DDT for malaria control purposes in Greece and Yugoslavia.[20]

Despite encouraging reports from German agriculturalists, the initial rejection of DDT by German health officials suggested that the "miracle" chemical was not universally celebrated.[21] The failure to develop a clear consensus on DDT may be attributed to the lack of coordination on the part of German scientists.[22] Nevertheless, the German experience with the new chemical illustrates the interwoven dimensions of the DDT story. DDT was not simply a chemical, but a compound that forced scientists to consider a myriad of issues connecting human health, insects, and the productive capacity of an agricultural system. These issues, moreover, were tightly bound by the concept of national health and were governed by a calculus determined by short-term and long-term risks and benefits. After the war, similar debates would emerge on a global scale that would revolve around different conceptions of public health, agricultural health, and environmental health.

For U.S. military and civilian health officials, however, the war erased these dilemmas. Of immediate concern was how to keep soldiers and civilians healthy in some of the most diseased environments in the world. This was particularly difficult since a majority of medical experts who came of age during the war did so without much knowledge of "tropical" medicine.[23] Of the various ailments that could affect Allied soldiers, malaria and typhus were the most widespread and dangerous. Though common in large parts of North and South America, Asia, Africa, and Europe throughout much of human history, at the start of the war, many in the United States considered malaria to be a tropical condition and typhus the problem of the "great unwashed." The realities of war created a steep learning curve for many who believed disease stopped at the border.

For Allied health officials, malaria and typhus created distinct problems. While both diseases are insect-borne afflictions, they encounter and infect the human body differently. As the war started, the epidemiology of both diseases was well known; the problem was how to stop them.

Transmitted by the bite of the anopheles mosquito, a malaria parasite — *Plasmodium falciparum* and *Plasmodium vivax* being the two most dangerous and debilitating — produces fever in the human host. The fever ranges from moderate to severe and can lead to blindness, cognitive impairment, coma, and even death. While victims of *P. vivax* may have recurrent fevers, those inflicted with *P. falciparum*, if fortunate enough to survive, will not experience periodic recurrences, unless another mosquito bites them. *P. falciparum* is the most deadly form of the disease and causes most malaria-related deaths in Africa today.[24]

The plasmodium connected humans to mosquitoes biologically and ecologically. Biologically, the plasmodium parasite requires both human and mosquito hosts in order to reproduce. The sexually reproductive stage of the parasite, called the gametocyte, circulates in the human bloodstream. When a female mosquito feeds on human blood, it picks up the gametocyte, which adheres to the midgut of the insect. There the gametocyte reproduces, eventually creating the infective stage of the parasite, called the sporozoite. When the female mosquito feeds again, the sporozoite enters the human body, infecting the person with the malaria plasmodium.

And because the insect vector varies widely, ranging from two to ten miles, malaria connects human bodies to environments that lay within and beyond the boundaries of human settlements, though these boundaries are never well defined. The ecologies of disease can be haphazard, as areas endemic with malaria are interspersed with "healthy" environments.[25] Rather than being simply a physical ailment, malaria broadens the terrain of disease connecting humans with nature, a connection that, historically, has been both profound and troubling.

Somewhat different, typhus is caused by another insect, the body louse. Lice infected with the parasite *Rickettsia prowazekii* (named after two medical researchers, Howard Taylor Ricketts and Stanislaus von Prowazek, who both died from the disease) feed on human hosts, depositing fecal material containing the parasite on the skin, material which then enters the bloodstream. Affected victims become nervous and restless, and may experience mental lapses. They also experience chills, fever, severe headaches, and spotted rashes. Serious cases lead to coma and death.[26] In 1909, Charles Nicolle, director of the Pasteur Institute in Tunis, confirmed that the human body louse transmitted typhus. (He would subsequently win the Nobel Prize for medicine in 1928.) Whereas malaria connected humans to a larger insect vector, the human body served as the host of body lice. So while unsanitary conditions contributed to the spread of the typhus, typhus control posed a different set of issues, especially as those issues related to the chemical control of the insect vector.

Although on the surface DDT seemed promising, there was still much unknown. How effective was it? What were the differences between Gesarol and Neocid? What insects did DDT kill? How should it be used? What were the potential hazards of such use? Could it be used to prevent the spread of typhus *and* malaria? "Our chief worry was," Knipling noted after the war, "can the chemical be used safely on man?"[27]

Knipling's laboratory conducted tests on insects and human subjects. Powder and spray formulas of DDT were applied in different concentrations to determine the most effective means of killing insects. Sprayed on walls, DDT was found to be a potent killer of mosquitoes since it remained active far longer than either pyrethrum or Paris green. In powder form, DDT remained lethal to lice four times longer than pyrethrum. "Test[s] against lice," Knipling reported, "fully confirmed the claim made by [Geigy] that DDT was a promising insecticide."[28]

Not only did DDT produce immediate results, but researchers were amazed at the compound's long-term toxicity: "The outstanding feature of DDT is its long-lasting quality."[29] Tests also revealed DDT's remarkable broad-spectrum properties. While it was effective against lice and mosquitoes, it also proved lethal to a number of insects, including bees, ladybugs, and other insects beneficial to agriculture and ecological balance.[30]

Experiments carried out on human subjects confirmed that the chemical protected the body from insect pests without visible signs of poisoning. In some cases, test subjects experienced minor irritations or rashes, but all signs indicated that DDT was not acutely poisonous to humans. Testing, however, was done in the context of war. Knipling wrote, "The application of control measures against adult mosquitoes in many aspects is even more important in time of war than during peacetime."[31] In wartime, the calculation of risk differed tremendously. Knipling's team was less concerned about the chronic toxicity of the chemical or its ecological hazards and more interested in determining the likelihood of acute toxicity. These experiments verified that under wartime conditions, DDT was safe in the short term. Less clear to Knipling and his team were the long-term impacts of the pesticide.

Despite these scientific breakthroughs, Knipling realized the war was not fought in a laboratory. So he, along with Percy N. Annand, director of the Bureau of Entomology and Plant Quarantine in Washington, enlisted a number of government and civilian research groups to conduct field tests. Perhaps the most important of these groups was the Rockefeller Foundation. Established in 1913, the foundation relied on scientific knowledge and educational reform to solve age-old problems for the "well-being of mankind throughout the world." Prior to the war, the foundation funded medical and social science research aimed at alleviating public health problems within the United States and beyond its borders. The Rockefeller Foundation developed a series of landmark public health projects around the world under its International Health Division (IHD), including malaria research in Italy and Mexico and yellow fever research in Brazil. By the Second World War, the foundation had established itself as one of the leading philanthropic organizations in the world, as well as a recognized leader in the field of public health. Deeply committed to the practical use of scientific research, the foundation played a critical role in developing public health remedies for military and civilian populations. According to the foundation's annual report of 1944, the organization worked tirelessly on number of insect-borne disorders, including malaria, which had "been proclaimed the disease enemy number one by the Army and Navy of the United States."[32]

As early as 1939, Dr. Wilbur Sawyer, director of the foundation's IHD, began working with U.S. military leaders on typhus prevention methods. In 1940, the foundation established a small working group to conduct experiments with vaccines and insecticides in Mexico, where typhus was endemic. That same year, Sawyer visited the laboratory of Harvard bacteriologist and typhus expert Hans Zinsser. Sawyer and Zinsser agreed to establish a program under the auspices of the foundation's IHD with a laboratory located in New York City.

Initially, the project was slow to develop, but by 1942, the foundation had created the "louse lab," which joined forces with Knipling's Orlando laboratory soon after. The two research groups began "to pool both information and material," yet they both played to their strengths.[33] The Orlando group concentrated on synthesizing DDT and conducted experiments in controlled environments. The foundation, on the other hand, produced lice in its New York laboratory and conducted experiments "in the field." These field experiments determined whether the lab results could be reproduced in different environments. Between 1942 and 1944, the foundation used DDT as a louse powder on medical students in New York, conscientious objectors in New Hampshire, and civilian populations in Mexico.[34] These tests confirmed the remarkable nature of the chemical. DDT was lethal to insects, but "safe" for humans. Yet these field tests were not without their problems.

Unlike the experimental groups in New York and New Hampshire, all of whom agreed to the tests, Mexican civilians were often not made aware of the testing.[35] One report indicated that the delousing campaigns using chemical pesticides would be better if health officials "obtain[ed] the understanding and cooperation" of local populations, rather than impose strict regimentation.[36] While small in scale relative to the global campaign yet to come, the efforts in Mexico exposed a critical problem in the delivery of public health services: How to do it? For example, a foundation officer conducting malaria work in Mexico reported that residents of Temixco were "somewhat antagonistic to the work, because they are not receiving any benefits from it and do not readily permit observations to be made in their home."[37]

While the aim was to alleviate the potential health risks to local populations, the question remained: What should be the role of the public health provider to ensure compliance and cooperation with the scientific project? Does the consent or cooperation of local populations matter? Or does the authority derived from scientific expertise supersede the concerns of local peoples? And what of the understanding of chemical technologies? How did DDT, a "modern" weapon of war against insects, "inscribe" a model of public health onto a foreign landscape or body?[38] While these questions were not new—indeed, the history of experts subjecting unwilling people to modern medicine is regrettably long—they remained difficult to answer. The global war compounded these complex issues since it projected American ideas about health, science, and chemical technologies onto people from different countries and cultures.[39]

These problems were perhaps compounded by the appointment of the pre-eminent malariologist Fred Soper to Roosevelt's new typhus commission in 1943. Since 1928, Soper had led a number of public health campaigns in Latin America under the auspices of the Rockefeller Foundation's IHD. At the time of his military commission, Soper had just completed a mosquito eradication campaign in Brazil that ended an outbreak of yellow fever, a debilitating viral disease that left its victims jaundiced, feverish, and nauseous, and in serious cases led to coma or death. Soper's campaign, while remarkable in its own right, was built on the work of others.

In 1881, the Cuban physician and scientist Carlos Finlay established that the yellow fever virus was transmitted by mosquitoes, rather than through human contact. Subsequent research found that the principal insect vector responsible for yellow fever was the mosquito *Aedes aegypti*. His discovery encouraged public health officials to rethink their approach to yellow fever. Rather than attack the symptoms in human sufferers, Finlay and other health officials considered mosquito control programs the most viable form of disease prevention. Finlay's work would later be confirmed by the American doctor Walter Reed, who witnessed the danger of yellow fever while surveying U.S. encampments during the Spanish-American War in 1898. In 1900, Reed returned to Cuba as the head of the U.S. Army Yellow Fever Commission for which he conducted numerous experiments, including tests on human subjects, many of whom served directly under Reed. Because of the epidemiological work of the commission, Reed became the celebrated figure of yellow fever control and fundamentally transformed the ways in which public health officials treated yellow fever.

While nineteenth-century public health officials held that the disease could be controlled by sanitation and quarantine, believing germs were spread by human contact and travel, Finlay and Reed argued otherwise. Reed's experiments, historian Margaret Humphreys argues, made the system of quarantine "foolish" and "obsolete."[40] Indeed, the elimination of yellow fever depended on mosquito control, not quarantine or sanitation. And because the insect vector was the root of the problem, environmental control, rather than curing the diseased body, seemed like the most effective and efficient way to suppress yellow fever.

Arriving in Brazil in 1928, Soper developed a single-minded commitment to mosquito control as the most effective means to combat yellow fever. His arrival coincided with reports of sporadic yellow fever outbreaks, even though Brazilian officials believed the disruption was firmly under control. In fact, some Brazilian health officials proposed the suspension of the yellow fever control program. A dedicated yet hardheaded administrator, Soper challenged Brazilian public health officials to develop a more rigorous record-keeping system and establish a vector control program. Soper maintained that Brazilian health officials failed to adequately document cases of yellow fever, since symptoms often resemble those of the flu. Committed to a strategy of chemical controls for the mosquito vector, Soper instilled a highly disciplined approach to public health, revolutionizing the delivery of health care systems. Soper "wrote the book" on yellow fever control.[41] Along with Soper's mosquito control program, researchers at the Rockefeller Institute developed a yellow fever vaccine in 1937 that was widely distributed in Brazil the following year. These developments underscored the foundation's multi-tiered approach to public health, a system that combined laboratory research with fieldwork, scientific knowledge deployed with dispassionate efficiency, and the unquestioned belief in technological solutions to complex ecological problems.

Following his yellow fever work, Soper directed a successful campaign against the malaria vector—Anopheles gambiae—in Brazil. An imported pest from Africa, the gambiae proved dangerous and elusive. By the 1930s, the mosquito had infected nearly 30,000 Brazilians with the malaria parasite, causing nearly 10,000 deaths. Environmental controls provided some protection against the mosquito but failed to eliminate the threat completely. Soper, however, embarked on a transformative program not only to control mosquitoes, but to eradicate them. In describing the campaign, Soper later recalled that "the method used was a straightforward chemical attack with paris green."[42] By 1938, Soper claimed victory because A. gambiae, though not the plasmodium, had been driven from Brazil.[43] As he had in his work on yellow fever, Soper developed a method of public health that would revolutionize—for better or worse—health delivery systems in the postwar years. By constructing a comprehensive attack on mosquitoes, including detailed record keeping, efficient deployment of manpower, and chemical controls,

Soper introduced the concept of "malaria discipline" into the public health lexicon. His method involved regimented science, deployed with conviction of purpose and commitment to health. But it also reduced local populations to passive actors or, even worse, belligerent distracters, while undermining local knowledge about insects, sickness, and health.

Soper brought his unique administrative skills to the U.S. Typhus Commission. Yet war created a unique set of challenges for public health officials that the landscapes of disease Soper faced in Brazil had not. As Reed and Soper previously discovered, in the best conditions, disease was difficult to manage and control. The transformation of wartime environments—infrastructure damage, the lack of clean water, and poor sanitation conditions—further exacerbated the problem. From a military perspective, malaria created a barrier that separated diseased and healthy bodies, just as yellow fever and typhus had, a barrier between those willing and able to fight and those incapable of doing so.

Despite these environmental conditions, war helped officials administer public health programs to soldiers and civilians alike. Because of the rigid structure of the military, public health officials could impose disciplined healthcare strategies on populations without their consent. Soldiers often did not have any control over their bodies. In some cases, soldiers were unknowingly exposed to dangerous substances in human experimentation trials.[44] Military health officials attempted to impose tight controls over human bodies and the environment. The Typhus Commission and the Malaria Control in War Areas programs, for example, granted public health officials and medical personal wide latitude over war areas and the people who inhabited them.

In July 1943, a team of Rockefeller officials visited the Orlando laboratory. Reporting back to IHD director Wilbur Sawyer, Raymond Shannon, a member of the Rockefeller team, claimed, "Mr. Knipling and his 20 cohorts showed me the 'magic' of their new insecticide, Gesarol." The compound, Shannon remarked, could "eventually lead to the extermination of such superdomestic anopheline as *gambiae* and *darlingi* and *Aedes aegypti*."[45] Shannon's enthusiasm over the promise of Gesarol expressed a dramatic turnabout in the idea of insect eradication, an approach that would come to dominate public health ventures in the postwar world.

Just as they had in their joint work on typhus, Knipling's Orlando laboratory and the Rockefeller Foundation conducted numerous experiments with DDT. But in this case, the experimental landscape widened, as work was done in the United States, Mexico, Trinidad, Brazil, and Egypt.[46] The Office of Malaria Control in War Areas, operating under the Public Health

Service, conducted much of the domestic fieldwork. Expanding the testing areas beyond the nation's borders did more than simply underscore the global dimensions of the malaria problem; it also acknowledged the realities of ecological difference. Ecological variety complicated the wants of military leaders and public health experts to control diseases across the battlefields of the Second World War. Environmental conditions in and around an army base in Arkansas, for example, were completely different from those in Algiers, Egypt, or the South Pacific. So whether abroad or at home, the particularities of the natural world required entomologists and public health experts to rethink their approach to field testing. One Rockefeller Foundation official suggested, "The Gulf Coast of Mexico might be of interest . . . because of the tropical conditions offered."[47] Similarly, Trinidad and Brazil offered U.S. health officials environments that were similar to the tropical areas where American servicemen and -women were being deployed.[48] Conducting field tests far and wide provided American entomologists with a better understanding of how DDT might perform on the battlefields of the Mediterranean and the South Pacific.

Soper was at the center of the most important field tests in the Mediterranean. While officially assigned to the U.S. Typhus Commission, he continued to work on the malaria problem under the Rockefeller Foundation's IHD. Traveling through Algeria, Morocco, and Egypt, Soper conducted surveys of mosquito populations, developed control programs, and began working with a new compound few had ever heard of.

Arriving in Egypt in January 1943, Soper quickly embarked on a myriad of projects—he was not one to waste time. Fresh off his eradication campaign in Brazil, he would once again confront the *gambiae*, which had recently "invaded" the Nile River Basin from the south. Previously unknown to the region, the *gambiae* carried a more lethal form of the malaria parasite—*Plasmodium falciparum*—placing local populations at risk since very few people had developed resistance to this strain of the disease. Soper advised British and Egyptian leaders to launch an eradication campaign using Paris green, a campaign modeled on Soper's work on yellow fever. But due to wartime shortages and the general indifference of British military leaders regarding the viability of eradication, Soper was not asked to oversee the project. Instead, at the request of British officials, Egyptian health officials undertook the project themselves, with disastrous results. Soper later recalled that the "failure was particularly frustrating because of my conviction that the solution to the *gambiae* problem was known, and I had had an opportunity to discuss it at a high level with all interested groups."[49]

According to reports, nearly 750,000 people contracted the disease between 1942 and 1945, causing the death of between 100,000 and 200,000 people. Although at the time the malaria outbreak was seen as simply a natural problem, multiple factors, many of which were man-made, contributed to the outbreak of the disease. The radical transformation of the Nile River Valley created environments suitable for the *gambiae* to reproduce and malaria to spread. Dam building, the construction of diversion canals, and the adoption of modern agricultural techniques, including the use of chemical fertilizers, recast the river valley into a highly productive environment, but in doing so, eliminated natural barriers that protected humans as natural defenses against *gambiae*. During the war, military travel up and down the Nile also contributed to the spread of disease, as the *gambiae*, a mosquito species with only a two-mile flying radius, traveled along with military boats and light aircraft. As the mosquito and the plasmodium spread north, Egyptian health officials, with very little assistance from their British counterparts, attempted to resolve the situation themselves by using a combination of chemical and nonchemical treatments.[50]

Unable to manage the spread of the *gambiae*, British officials reluctantly accepted the Rockefeller Foundation's assistance and the appointment of a "malaria dictator." For many reasons, professional and personal, Soper was the obvious choice for the post. In May 1944, Soper established an eradication project based on his Brazilian program that would eradiate the *gambiae* from the Nile River Valley by 1945. Only now DDT was his chemical of choice.[51]

Soper, however, had not been working exclusively on the Egyptian malaria problem. Frustrated by the lack of coordination and the failure of Egyptian leaders to accept the Rockefeller Foundation's assistance early in the malaria outbreak, in the spring of 1943, Soper traveled west to Algeria. There he worked on the typhus problem with the American Red Cross and the Pasteur Institute, a French research institution directed by Dr. Edmond Sergent. For Soper, a committed scientist who believed in the power of technology, Algeria was a perfect place to develop strategies to eliminate typhus. Not only did he have the financial and intellectual resources around him, but he also had a captive population in which to conduct experiments—prisoners.

Soon after his arrival in Algeria, Soper began conducting tests of two experimental delousing powders designated GNB (barium sulphate) and MYL, a pyrethrum louse powder, on prisoners. The captive population served as an important testing ground for both the chemical and the process of application. And while the chemical powders demonstrated significant results, leading Soper to claim that "two treatments of either GNB or MYL can hold a louse population in chemical for 2 months," it was the process of applying the chemicals to human bodies that offered public health officials added encouragement. Soper recalled, "The prisoners were anxious to be powdered and were most helpful in handling clothing and blankets and even in operation of the dusting machines."[52]

As tests moved from prisoners to civilians, Soper noted in a letter to Brigadier General Leon Fox, director of the Typhus Commission, that "those who know the local mores insist that it will be completely impossible to powder the native population by removal of all garments. To meet this situation . . . initial tests of an agricultural dusting machine were made."[53] Dusting with hand sprayers, Soper discovered, offered the most effective means of applying delousing powder to people unwilling to disrobe because of cultural or religious beliefs. Here, unlike in his malaria work in which the *Anopheles gambiae* could be controlled away from the body, the typhus louse posed a set of problems that connected human and insect bodies. Because the problems of malaria and typhus were quite different, these tests were not simply based on the chemicals' effectiveness against insects; they had to be safe for humans as well.[54]

But DDT was still only on the horizon. In March 1943, George Strode of the Rockefeller Foundation's IHD, who during the war worked for the State Department and the Red Cross, remarked while traveling in North Africa, "No one here ever heard of gesarol."[55] Within a few short months this would change, as DDT would become the most important insect control agent for the remainder of the war.

All of this, however, set the stage for the deployment of DDT on a massive scale. While early tests in Orlando identified the risk of human toxicity as real, researchers concluded, "The information available from all sources indicates that the proposed uses of DDT present little, if any, hazard if the recommended precautions are followed."[56] Soper's field tests also confirmed these findings. And the real test of the pesticide would come following the allied victory against German forces in Italy.

In the fall of 1943, U.S. forces defeated German troops in Naples and gained control of a city destroyed by months of intense fighting. There was little or no running water, gas, or electricity, and the civilian population was ill equipped to rebuild the city's infrastructure. As the *New York Times* reported, "Naples is suffering terribly from the depredations of the Germans, lack of water and food and spread of disease."[57]

In September 1943, Allied military health officials recorded nineteen cases of typhus in Naples. By October, the number of cases increased to twenty-five. In November, the number rose to forty-two, and by the end of December, a full-blown epidemic ensued, with 341 new cases reported.[58] The threat of typhus loomed large in the civilian population. Living conditions within the city created an ideal environment for spreading typhus since Neapolitans sought shelter from the war in the city's underground caves, where sanitation and running water were nonexistent. Official reports suggested that upward of 90 percent of the population carried lice.[59]

The Allied military government, fearful of the effects of a full-scale typhus epidemic on the war effort, and unable to combat the outbreak themselves, appealed to the Rockefeller Foundation's Health Commission. Despite his earlier decommissioning, Soper took the lead, much to the dismay of other military officials. By the end of December 1943, the Health Commission had established five delousing stations throughout the city and began a massive campaign of dusting civilians with DDT. Brigadier General Leon Fox also led eight medical teams and, with the assistance of the Rockefeller Foundation, organized a total of forty-two delousing stations. The "Second Battle of Naples" operated with military precision, as had the military battle before it, and soon brought the epidemic under control. By March 1944, the battle was all but over. In total, 1.3 million people were dusted with DDT powder at a rate of 50,000 per day. Fox claimed, "We dusted practically all of Naples. . . . Never before in all history has a typhus epidemic been arrested in midwinter. It's not a question of how rapidly we gained control—the important point is that we stopped it at all."[60] The *New York Times* reported, "Neapolitans are now throwing DDT at brides instead of rice."[61]

Soper responded to the remarkable turn of events. "It really came as a surprise to some of us," he quipped, "to learn that Egyptians, Arabs, and Neapolitans do not like being lousy and will make an effort to get rid of their parasites if given the opportunity to do so."[62] Both sarcastic and condescending, Soper nevertheless witnessed the widespread enthusiasm generated by the new pesticide.

While typhus had been stopped in Italy, malaria continued to threaten Allied forces.[63] Malaria rates at the start of the war were eight to ten times higher than battle casualty rates. Throughout the Mediterranean, malaria endangered soldiers. But the problem was worse in the Pacific theater. One report made after the first offensive on the Solomon Islands indicated that the Allies suffered 10,635 casualties, of which 1,472 were the results of gunshot wounds, while 5,749 were hospitalizations due to malaria.[64] The renowned malariologist Paul Russell later recalled that General MacArthur was not at all worried about defeating the Japanese, "but he was greatly concerned about the failure up to that time to defeat the Anopheles mosquito."[65]

Building on Soper's regimented mosquito campaign in Brazil, although much less committed to chemical controls than Soper, Russell promoted the concept of "malaria discipline."[66] He believed that in order to create healthy environments in which mosquitoes were a nuisance rather than a human health hazard, forceful leadership was required. In writing about his experiences in the South Pacific, Russell remarked, "The principal general lesson was that it is impossible to control malaria effectively in military forces in highly malarious areas unless commanding officers from the highest to lowest echelons are malaria conscious." He further stated, "A malaria policy must not only be formulated: it must be enforced. Malaria discipline is absolutely necessary to an army's success in fighting the *Plasmodium*-mosquito axis."[67]

Drainage ditches were constructed in an attempt to reduce mosquito breeding grounds. Synthetic antimalarial drugs, like Atabrine, served as therapeutic remedies for those who contracted the disease. Bed nets and insect repellents provided additional protection against mosquitoes. The military produced printed material, including posters, matchbook covers, and calendars, reminding all military personal to practice malaria discipline.[68] Yet underscoring, or in fact superseding, these methods was DDT, which quickly emerged as the primary weapon against insect pests.

Continued military development of DDT throughout the war created new techniques for delivery. Spray techniques improved as the military experimented continually with various solutions and methods.[69] In November 1944, *Science News Letter* reported that whole islands could be sprayed because of a "makeshift gadget" that enabled the release of DDT through a pinhole nozzle, meaning that the pesticide could be used on an enormous scale. Spray planes saturated whole islands in the Pacific.[70] Aerial applications of a petroleum-DDT mixture cleared mosquitoes and other insects from beachheads prior to Allied invasions, providing an insect-free area for advancing troops. Major General Norman T. Kirk, surgeon general of the army, reported to the Association of Military Surgeons that incidences of malaria in the army had been reduced by 75 percent due to "proper malaria discipline and control measures." British and Australian forces also deployed DDT throughout the South Pacific.[71] After the invasion of Saipan, an epidemic of dengue fever broke out, which, according to Major General Kirk, "was a mass of flies and mosquitoes on D-Day."[72] Massive aerial spraying with DDT quickly reduced the number of new cases by more than 80 percent in two weeks.[73] As a result, rates of all insect-borne diseases plummeted. According to one report, malaria rates in the South Pacific dropped by 95 percent in 1944 as a result of "scientific medicine and military discipline."[74] By the war's end, the military's Malaria Control Program likened DDT to the atomic bomb.[75]

Although DDT use increased during the war, the increase did not come without considerable risk. Despite the extensive tests conducted in Orlando and in the field by the Rockefeller Foundation, DDT was still a relatively unknown technology. Increasingly, the risks of DDT were becoming more evident. Doctors at the Public Health Service reported, "The toxicity of DDT combined with its cumulative action and absorbability from the skin places a definite health hazard on its use."[76] Brigadier General James S. Simmons remarked, "The preliminary safety tests, made with full-strength DDT, had been somewhat alarming."[77] In a joint statement about DDT's toxicity, Dr. H. O. Calvery of the Food and Drug Administration and Dr. Paul A. Neel from the National Institutes of Health observed, "The extensive animal experimentation and investigative agricultural uses indicate quite clearly, we believe, that DDT is a deleterious substance." Another Rockefeller Foundation official claimed, "In large doses it is toxic to fish."[78] Even one of the chemical industry's leading publications, Soap and Sanitary Chemicals, warned about the "health hazards of DDT."[79]

As early as August 1944, Time issued a warning, stating that "the extraordinary new insecticide DDT is not without drawbacks" and "may be toxic to people and animals as well as insects."[80] Other negative press reports followed. The Christian Science Monitor informed its readers of "DDT Dangers." A New York Times headline stated, "DDT Spray Called Injurious to Birds." Audubon "cautioned" its readers about the pesticide. Better Homes and Gardens warned, "DDT: Handle with Care." And the Ladies Home Journal declared, "DDT will get 'em, but . . ."[81]

Research on DDT's biological effects also intensified. In 1945, researchers at the national wildlife refuge in Patuxent, Maryland, conducted extensive tests to determine the ecological consequences of massive DDT spraying. Increased concern arose over DDT's impact on the "balance of nature" because of the chemical's broad-spectrum characteristics. In August 1945, the Christian Science Monitor claimed: "Like the atomic bomb, DDT eliminates ALL insects without regard to their destructive or beneficial habits."[82] That same year, a massive fish kill that biologists attributed to DDT sounded alarms as well. Clarence Cottam of the U.S. Fish and Wildlife Service warned against the indiscriminate use of DDT because "it is capable of considerable damage to wildlife, beneficial insects and indirectly to crops."[83] In a letter to the Rockefeller Foundation's president, Raymond Fosdick, Arthur Sulzberger, the editor of the *New York Times*, asked if the foundation could "look into the question of the effect of DDT on animal life generally." Sulzberger's reasoning was unscientific, but ecologically sound. Sulzberger reflected, "It kills flies and mosquitos; surely then it must have some effect on other insects and, in turn, on birds."[84]

While initial research foreshadowed the problems of chronic toxicity and ecological destruction, within the context of war, risk was calculated differently. Public health officials and entomologists were more concerned about acute toxicity and the immediate impacts of DDT's use than about long-term consequences. Officials and entomologists believed their job was to help win the war, not make long-term decisions regarding the ecological safety of pesticides. As one medical poster read, "Shorten the War, Prevent Malaria." So for public health officials like Soper and Russell, entomologists like Knipling, and military commanders like MacArthur and Fox, the risk was determined to be acceptable.[85]

Much of this nuance was lost on U.S. chemical makers, who, on the one hand, wanted to share in the credit for the public health successes of the war, and, on the other, wanted to create demand for a postwar domestic market. The president of the National Association of Insecticide and Disinfectant Manufacturers, Henry Nelson, proclaimed, "Ever since the start of World War II, our industry has been engaged in work vital to the protection of the health of our armed forces."[86] Another commentator argued, "The war demands were urgent and industry responded promptly."[87] Indeed, by war's end, U.S. manufacturers produced 36 million pounds of DDT, up from the 153,000 pounds produced only two year earlier.[88] In August 1945, when the War Productions Board lifted the wartime restrictions on the domestic consumption of DDT, the chemical industry proudly proclaimed, "Here it is!" and "Released for Home-Front Use." One advertisement remarked, "DDT had a good Publicity Agent."[89]

SOON AFTER THE WAR, James Phinney Baxter, the official historian of the Office of Scientific Research and Development, remarked on the influence that events overseas would have on the United States in the next few years. He wrote, "It is interesting to observe that out of this rather frenzied search for means of killing lice in Italy and mosquitoes in Assam have come substances that will affect our own country, in which lice and mosquitoes have only a nuisance value."[90] Interesting indeed; the public health war fought in multiple arenas and across distant lands had a profound impact on the United States. The public health effort changed the ways in which Americans understood insects and environments. At first, DDT seemed to have the potential to rid the world of a number of insect-borne diseases. But public health successes made insect control deceptively easy. In the years to come, some of the lessons of the war would be forgotten and others would be embraced, with renewed faith in the power of human technologies.

For their part, Knipling, Soper, and Russell were leading actors in a network connecting government institutions, scientific foundations, and the chemical industry. It was a world increasingly interconnected. Speaking before the Senate Subcommittee on Wartime Health and Education in December 1944, Brigadier General Simmons proclaimed, "Within the short space of four years these joint researches in the laboratory and in the field have afforded greatly improved methods for the control of a number of diseases."[91] Reiterating this idea, Geigy's Victor Froelicher remarked, "Too much credit cannot be given the Orlando station for the amazing developments of military uses for DDT composition."[92] George Strode, of the Rockefeller Foundation, also chimed in, "I know of no group which worked more whole-heartedly, effectively and without thought of personal benefit than the Orlando group."[93] These sentiments would not end with the war's conclusion. Rather they would continue to define the usage of DDT within and beyond the nation's borders.

The importance of insect control during the war was not lost on the editors of the *Journal of Economic Entomology*. The duration of the war was "the Golden Age of Entomology," they claimed. "Never in the history of this science has it risen to such heights of importance and recognition." Entomologists, they argued, now stood "shoulder to shoulder with medical doctors, the sanitary engineers, and also with the combat officers and soldiers." In many ways, it was a new world, one in which entomologists deployed a new chemical weapon, DDT, with military efficiency. How they would use this new weapon was unclear, but the war established specific models to be used in the future.[94]

The scientific networks developed during the war made a forceful argument for American intervention in the postwar years. New instruments of disease control confirmed in the minds of many American health officials that insect-borne infections could be brought under control, perhaps even eradicated. Even the preeminent scientist Vannevar Bush understood this all too well. Summarizing the conclusion of his influential report from 1945, "Science: The Endless Frontier," Bush wrote in the *Atlantic*:

Of what lasting benefit has been man's use of science and of the new instruments which his research brought into existence? First, they have increased his control of his material environment. They have improved his food, his clothing, his shelter; they have increased his security and released him partly from the bondage of bare existence. They have given him increased knowledge of his own biological processes so that he has had a progressive freedom from disease and an increased span of life. They are illuminating the interactions of his physiological and psychological functions, giving the promise of an improved mental health.[95]

For Bush, science offered the promise of a better world. And DDT, a chemical that made ameliorating disease on a global scale imaginable, was certainly one of the advances that could ensure a better world. Winning the war also positioned American science in an enviable position to promote global health while expanding the nation's interests. In seeing this opportunity, Bush shared much with Henry Luce. Both men believed in U.S. power as a force for good in the world. They both had an unyielding faith in science and technology. And they firmly believed in the inherent goodness of the United States. And while Bush and Luce said very little about DDT publicly, their shared vision of U.S. power and global influence placed the chemical squarely at the forefront of the burgeoning American Century.

Not to be forgotten, however, is the fact that the United States improved public health prior to the war without DDT. Yet by 1945, public health experts believed DDT could vanquish insect-borne disease singlehandedly. And once again, Italy would be at the center of another massive DDT experiment. This time the effort was not just about annihilating an insect; it was about modernizing underdeveloped economies and stifling the threat of Cold War communism. Yet, just as the fallout from the atomic bomb had unintended consequences, the use of DDT called into question the very scientific innovation that created it in the first place.

2

DISEASE, DDT, AND DEVELOPMENT

The American Century in Italy

*Life looked at what people thought and did in a particular way. It stood
for certain things, it entered at once the world-wide battle for men's minds.*
—Wilson Hicks, *Life's* photo editor, 1952

*The results of successful demonstrations in Sardinia could be
applied with the necessary modifications in other areas.*
—Lord John Boyd-Orr, Nobel laureate, 1950

In 1934, the editors of *Fortune*, the nation's premier business magazine, devoted the entire July issue to Italy. For Henry Luce, *Fortune's* founding editor,
Italy's fascist regime both intrigued and repulsed him. The sheer force of
Benito Mussolini's political will to radically transform and modernize the
Italian state reflected Luce's search for political and professional stability
and order at home. "The good journalist must recognize in Fascism certain
ancient virtues," the magazine acknowledged in an introduction that was
unsigned, though likely written by Luce. These virtues included "Discipline,
Duty, Courage, Glory, [and] Sacrifice." At the same time, the antidemocratic
impulses of Italian fascism gravely concerned Luce. *Fortune's* Italy issue,
therefore, was an attempt to "look at the creature from several angles."[1]

Much of the issue described the structural changes to Italy's political economy, with articles focusing on a range of topics from "fascist finance" and
the "corporative state" to "agriculture and industry." Other articles focused
on the church, the legacy of the Italian nobility, and Italy's "colonial empire."
Fortune also profiled Mussolini, portraying him as a devoted family man and,
not surprisingly, a strong-willed leader.

Another underlying theme of the issue was health. Mussolini's quest to raise a new generation of strong, able-bodied fascists engendered improvements to early childhood health and calls for higher birth rates. "He is making millions of Italians healthy and robust," *Fortune* pronounced. The magazine also praised Mussolini's reclamation projects, which converted once dangerous swamplands into productive farmlands. "A tremendous Fascist to-do," *Fortune* reported, "surrounds projects like the draining of the Pontine Marshes," a malaria-infested region south of Rome that had plagued inhabitants and travelers for centuries.

Luce continued to pay an extraordinary amount of attention to Italy in the years after 1934. Not only did the fascination with fascism guide his intellectual pursuits, but the outbreak of war and the geopolitical struggle in its aftermath placed Italy squarely in the center of world events. As a newsman, Luce was committed to bringing the story to the American public through his expanding magazine empire that, by 1936, included *Time, Fortune,* and the highly popular *Life.* Italy figured prominently in the pages of Luce's magazines, just as much of war-torn Europe did.

Luce's interest in Italy was intellectual and economic, but also personal. His estranged wife, Clare Booth Luce, served as the U.S. ambassador to Italy during the Eisenhower administration. Italy seemed to offer some solace to the Luces in their troubled marriage, and Henry would travel frequently to Rome. There he socialized with political dignitaries, conducted business, and witnessed firsthand the escalating tensions of the Cold War.[2]

As a publisher, business tycoon, and writer, Luce reflected on and shaped the world around him. His influential essay "The American Century" became a defining document of its day, as it recognized the nation's global importance while challenging the history of American parochialism.[3] Indeed, his vision of the American Century encouraged business leaders, philanthropic organizations, and federal agencies to spread their influence around the world. Americans, Luce argued, must "accept wholeheartedly our duty and our opportunity as the most powerful and vital nation in the world and in consequence to exert upon the world the full impact of our influence, for such purposes as we see fit and by such means as we see fit."[4]

His magazines also provided Americans a lens through which to examine and understand the transformative events of the twentieth century, including the rise of fascism, the Second World War, and the Cold War. *Life* quickly established itself as the nation's leading weekly magazine, with a wartime readership of approximately 20 million. Between 1936 and 1953, Luce's magazine empire depicted a tumultuous world, one in which health and stability were under constant assault. And in keeping with his belief in American exceptionalism, Luce projected a vision through his magazines' pages of the United States as the necessary guarantor of stability, whether in the realms of economy, politics, or health.

In its own way, DDT was part of this larger narrative of American expansionism and internationalist gaze. The chemical that single-handedly reversed the typhus epidemic in Naples and offered much-needed protection against the anopheles mosquito appeared sporadically in the pages of Luce's magazines. When it did, DDT became a powerful symbol of American wartime and postwar aspirations. The pesticide was never far removed from the foreign policy ambitions of Luce and like-minded internationalists.

Nowhere was this more pronounced than in Italy, which became the focal point for two massive DDT experiments. Naples, as we have seen, was the location for one such trial. The other, a malaria eradication project, took place on the island of Sardinia. American-led public health programs deployed DDT to improve human health and encourage economic growth, much as Mussolini's reclamation projects had attempted to control nature for human consumption.

An understanding of the ways these stories were told to the American public is important to understanding the history of DDT. Throughout the war and postwar period, Americans were presented with a series of highly scripted narratives about Italian health. Individually, each story, image, or film conveyed specific ideas about American technology and humanitarianism. Collectively, they captured something more essential about U.S. power and the multiple meanings of DDT during and after the war. While other news sources reported on the transformative events that unfolded in Italy—some of which are documented in this book—Luce's magazine empire told a particular story about DDT and its impacts in Italy.

SOUTH OF ROME LIES THE APPIAN WAY, an ancient road carved through the Italian countryside. Designed to quickly move Roman troops throughout the southern peninsula, early sections of the road date to 312 B.C.E. The section connecting Rome with Capua, an important crossroads of the Roman Empire, went through the Pontine Marshes, a low-lying swampland that was one of the most malarial regions in Italy. For centuries the Pontine Marshes threatened travelers and stifled many plans to settle the region. Stagnant pools of water and warm climates made it a perfect breeding ground for the *Anopheles labranchiae*.

To Mussolini, the Pontine Marshes represented a challenge. Transforming the malarial region into a thriving agricultural sector would validate the modernizing impulses of his fascist regime. In 1928, the Italian government established "Mussolini's Law." It was part of a development scheme called *bonifica integrale*, a three-part plan to drain marshes, build homes and establish farms, and protect new settlers from the *labranchiae* through health services and by screening houses.[5] Mussolini's law viewed nature simply as an impediment to modernity. Nature was to be conquered, transformed, and controlled.

By the time Henry Luce decided to examine Italian fascism in 1934, *bonifica integrale* was an overwhelming success. The marshes had been drained by a series of technically advanced pumping stations that diverted water into a manmade canal system. Houses and towns were constructed. In 1932, Mussolini established the provincial capital of Littoria (now Latina) in the heart of the Pontine Marshes as a shining symbol of his fascist regime. Farmers cleared lands for wheat production. "The peasant," *Fortune* reported, saw "the marshes of the Campagna [an area south of Naples] finally drained after 2,000 years of puttering, and their fever-ridden thousands of acres converted into good farm land."[6] It was development writ large. Yet it did not completely solve the malaria problem, and the marsh would not stay drained for long.

Despite Mussolini's best efforts, malaria "rage[d] continuously" in 1936 in certain parts of the marshes, while other areas witnessed a steep decline in transmission. The Second World War also destroyed any gains made in malaria prevention in the 1930s. Malaria rates in Italy skyrocketed as a result of the war. Littoria, once the radiant representation of Mussolini's modernization efforts, "suffered most severely from the war-induced epidemic."[7]

But perhaps the most devastating effect of the war was the purposeful destruction of the pumping stations by withdrawing Nazi forces. German bitterness about Italy's political reversal in 1943, a change from ally to enemy, unleashed a torrent of violence throughout the southern peninsula. "Occupation," historian Frank Snowden wrote, "rapidly produced a crescendo of violence as the Germans embarked upon the tasks of plundering Italian economic resources, conscripting forced labor, destroying Italian Jews, and terrorizing the population into submission."[8]

To slow down Allied troops, German soldiers dismantled much of the infrastructure sustaining the reclamation project, destroying bridges and canals and sabotaging the pumping stations. Soon the region was once again underwater. Travel through the region became difficult, if not impossible, and malaria came back with a vengeance.[9]

Not surprisingly, Luce's brief infatuation with Italian fascism ended as the war began. No longer did he or his magazines hold Mussolini in such high regard. *Life* even ridiculed a project Luce once praised. Referring to Italy's attempt to become self-sufficient in food production, the magazine reported, "It was a costly and retrogressive effort, resulting in such economic fallacies as Mussolini's campaign to grow wheat on the Pontine Marshes (some of the most expensive wheat ever produced)."[10] War, of course, shaped *Life*'s characterization of the fascist project. But when the United States began spreading the gospel and technology of development, criticism of such "costly" and "retrogressive" projects soon vanished from most U.S. publications, including *Life*.

Luce's magazine empire provided dramatic coverage of the war. In issue after issue, his magazines transported readers to distant places, from the South Pacific to North Africa, from the shores of France to the Eastern Front. *Life*, in particular, captured the imagination of its readers, primarily due to its use of photographs. Indeed, Luce created a visual medium that was clearly the most recognized source of photojournalism in the nation and, according to historian Wendy Kozol, "the primary vehicle for conveying the news visually to a mass audience."[11] Throughout the global conflict, *Life* became the preeminent venue for publishing images of war. Whether of ally or enemy, *Life*'s war photography provided Americans with a visual representation of a world beyond its border, creating a "visual economy" of war and health that fit neatly with the aspirations of the makers of U.S. foreign policy during and after the war.[12]

The importance of the Naples typhus campaign was not lost on *Life*'s editors. Certainly other news outlets reported on the massive public health campaign, but only *Life*, in what was a staple of the Luce publishing empire, captured and communicated the events in Naples through photographs. These images are a window into the complicated history of DDT and serve to underscore the politics of American health adventures in Italy.

Following the British and American assault on Nazi positions in Salerno and Naples, Bert Brandt, a staff photographer for United Press International and Acme Newsreel Pool, spent the early part of 1944 documenting the work of the Allied Typhus Commission in Naples. His images captured the organized chaos of the public health campaign and showed the connection between diseased bodies, Allied health personnel, and chemical technologies. While much of Brandt's work during the war focused on combat, including taking one of the first pictures of the Normandy invasion, perhaps his most striking images were not of war, but of health.[13]

Brandt's images highlighted the public health aspect of the war. Their publication in *Life* brought the campaign from the battle front to the home front, connecting American audiences to public health in Italy. And the chemical that made the typhus campaign a "success," DDT, reaffirmed the wartime benefits of science and technology. For a nation largely unaware of the scientific work being done by Knipling's Orlando laboratory or by Soper in his field experiments in Algeria and Egypt, Brandt's images not only introduced the nation to DDT, but also provided undisputed visual evidence of the power and potential of the "miracle pesticide."

These images represented more than the fulfillment of U.S. intervention in Italian public health, however. Unbeknownst to Brandt, they also symbolized the beginning of a longer process of exchange that transcended the war, connecting Italy to American ideas about health, economic development, and chemicals. His images would help Americans understand the military's efforts to alleviate pain and suffering in a foreign land.

Life, however, was not the only news outlet to report on the Naples typhus campaign. The *Saturday Evening Post* published an article by Allen Raymond, titled "Now We Can Lick Typhus," which described the connections between war and health.[14] "I knew that into whatever lands our armed forces marched with their engines of destruction," Raymond wrote, "right beside them would be our medical forces with their weapons of healing, adequate not only for the protection of our own but bent on aiding other peoples less fortunate." *Reader's Digest* recounted the story in its June 1944 edition. That same month, *Time* reported that DDT was "one of the great scientific discoveries of World War II." And *Newsweek* claimed it was "one of the three greatest medical discoveries to come out of the war."[15]

Life offered visual evidence of the typhus campaign that emphasized the interaction between public health officials, Italian civilians, and DDT. *Life* published eight of Brandt's photographs in a single issue, capturing various stages of the typhus campaign. Each photograph told a story about DDT, the typhus campaign, and U.S. interests abroad. Of the images, those *Life's* photo editors selected for publication focused almost exclusively on children, women, and the elderly as the primary benefactors of military aid.[16] Only two images showed military servicemen engaged in activities other than providing health services to women, children, or the elderly. One of them depicted soldiers using a cement mixer to mix DDT louse power. The other showed Italian military personal getting vaccinated. The remaining images portray representatives of the Typhus Commission with their Italian subjects, capturing the dynamic interaction between U.S. military workers and Italian bodies. Collectively, the images reflect older notions of poverty, health, and cleanliness.[17] They cast healthcare workers as saviors of the ill-clad, ill-fed, and

Bert Brandt photograph of a Rockefeller Foundation official looking for lice in the hair of a young Italian girl. Originally published in *Life*, February 1944. Corbis.

ill-housed sufferers of disease. One shows a Rockefeller Foundation official looking for lice in the hair of young Italian girl. Her mother and siblings surround her; her father is conspicuously absent, presumably because he is fighting in the war or has been a casualty of it. The family is unkempt, posed in sharp contrast with the uniformed official.

Another image plays off the same family dynamic, but in this case Brandt captures an infant at the forefront of the action. Here a public health officer delouses an infant while adult family members smile approvingly at the action unfolding before them. It all seems so innocuous in the gaze of Brandt's camera.

In making women and children the centerpiece of the essay, *Life* drew on themes evoking paternal notions of U.S. interests abroad. Brandt was not alone.

Fred Toelle, an American artist stationed in Italy with the Medical Museum and Arts Service, portrayed similar themes. As an illustrator and painter,

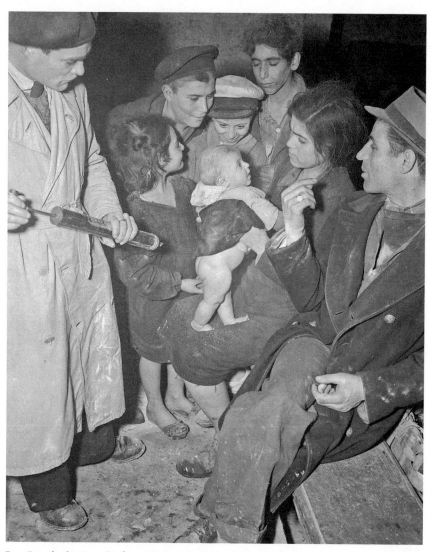

Bert Brandt photograph of a public health officer delousing an infant in Naples. Originally published in *Life*, February 1944. Corbis.

Toelle had considerably more time to construct his visual story of American public health efforts. Despite the inherent differences between photography and painting, Toelle's depictions of DDT-related projects in the Pontine Marshes were remarkably similar to Brandt's photographs. In a 1944 painting, for example, Toelle not only illustrated the process of indoor residual spraying, overseen here by three men in uniform, but the scene is set with a mother and child who happen to be standing directly under the spray gun.

Fred Toelle painting, *Home Disinfection, Pontine Marshes*. Courtesy of DeGolyer Library, Southern Methodist University, Dallas, Texas (Ag2002.1454), and the Toelle family.

Again, two themes emerge. First, American soldiers are protectors of Italian health. DDT, spray guns, and military labor offer safety and security to innocent victims of war and disease. Second, the recurring theme of mother and child is rife with obvious religious symbolism. While Italians were overwhelmingly Catholic, the mother and child motif also played on dominant U.S. religious beliefs. The trope was a symbol of commonality rather than difference, one that forged what has been described as an "Atlantic Community" between the United States and Italy.[18]

Concomitantly, these images express a type of "benevolent supremacy," an idea that the U.S. global presence was intrinsically different from the oppressive nature of European colonialism. Melani McAlister describes this reconceptualization of global power as "particularly meaningful *within* the United States, as a *self-representation* for Americans about their growing international power."[19] These symbolic expressions of health, technology, DDT, soldiers, and women and children created a visual landscape that confirmed the altruistic nature of U.S. power. More than anyone else, Luce had both the ability and means to promote and define the meaning of American expansionism along these lines.

His influential essay "The American Century" (1941) was one such expression of American expansionism. Luce understood that military power was but a small part of making the American Century. For Luce, ensuring human health, whether that involved eliminating disease or feeding the people of the world, was critical to forging the American Century. "Sickness of the world is also our sickness," Luce reminded his readers. And if the twentieth century "is to come to life in any nobility of health and vigor," then, Luce submitted, it "must be to a significant degree an American Century."[20] Brandt's images captured the essence of a political project that Luce envisioned. They captured beneficent engagement with the world beyond U.S. borders, engagement using American technology to improve the lives of others.

Yet Brandt's images depicted histories that were much more contested. One of the more powerful visual representations of the spraying of DDT in Naples portrays a woman's discomfort with a blast of DDT powder administered by a public health official. "Woman flinches," the caption reads, "as an Italian duster gives her a blast of the powder." The woman, lying in a prone position, suggesting that the disease may have already taken hold of her, turns away from a direct blast of DDT. This image evokes the violence of war, whereas other photographs use relatively calm imagery involving mothers and children. The public health official, presumably an Italian health worker, armed with his rifle and out to exterminate the enemy, calls to mind a foot soldier on the front lines who is forced to assesses risks and benefits much differently than military (or public health) planners do. In many ways this image captures the zeal with which the typhus campaign was carried out.

Collectively, the images captured and affirmed power relations that fit neatly with the ways the American public viewed itself and the world. These interpretations drew on familiar themes of U.S. foreign interventions and "reinforced common beliefs about technology, representation, and social structures."[21]

Indeed, the visual cues in these photographs link U.S. might with Italian helplessness, a relationship conveyed by distinct gendered characteristics that could be easily digested and understood within the United States. Neapolitans are cast as either helpless children or members of a nation aged and neutered by the ravages of war and disease. Images of Typhus Commission officials, conversely, emphasize the heroic aspect of their work and a paternalistic approach to public health.

The use of the spray gun heightened the military expectations of this pub-

Bert Brandt photograph of an Italian public health official spraying a woman. This image captures the zeal with which officials from the typhus campaign carried out their mission and suggests that some Neapolitans did not entirely embrace or accept the delousing program. Originally published in *Life*, February 1944. Corbis.

lic health campaign. Not only did the men of the Typhus Commission save lives and serve as surrogate fathers and husbands, but the spray gun also reaffirmed gender identities. "Our fighters," Hudson Spray Company, makers of a delousing spray gun, proclaimed, "need both kinds of guns."[22] Heroic, fearless, and fatherly, American public health officials used their scientific expertise to improve the lives of others less fortunate. They were strong, able-bodied men who could use technology proficiently. Saving the lives of women, children, and the infirm only magnified the heroic sense of the public health campaign in Naples, which created an almost mythical aura around DDT.

Despite the visible difference between disease and health, Brandt's photographs also captured a sense of shared humanity. Italian civilians were victims of war, first fooled by Mussolini's fascist promises and then injured by Nazi brutality. Essentially, because of this common connection, Italians were worthy of medical attention. Public health workers were simply doing their duty as part of an "Atlantic community," connecting Italian health to the technological wonders of American science. These depictions were in sharp contrast to images of the war in the Pacific.

Rather than illustrating a shared sense of humanity or an "imagined community," war photography in the Pacific captured racial differences. The Pacific war was indeed a "race war."[23] *Life* informed its readers, "The Japanese people are not actually inhuman. They behave with inhuman restraint or inhuman violence because they exist not as individuals but only as families, a nation, a race."[24] In confronting such an enemy, Americans launched an undeclared "total war" against the Japanese, with military and public health officials sharing the language of "annihilation" and "eradication" as a way to describe the violent conflict in the Pacific.[25]

The results of such a war left a completely different visual legacy than that of the war in Europe. Mutilated bodies of Japanese soldiers were common in the visual depiction of the war. Images of headless, mangled corpses juxtaposed against healthy Allied soldiers fixed the brutality of the Pacific conflict within a racialized discourse about the Pacific war. Indeed, graphic depictions of death were embedded in the visual economy of the Pacific in ways they were not in Europe. According to historian Susan Moeller, "No photographs of *caucasian* bodies *mutilated* by Allies were ever printed."[26]

Seen in this light, Brandt's photographs reaffirm the linkages of an Atlantic community between Italians and Americans during the war. But equal partners they were not. The deployment of American military, medical, and technological expertise to rid the Italian body of disease underscored the disparity in this relationship. Italians were simply recipients of American good will. These lessons would be replicated in the postwar period.

IN DECEMBER 1945, *Soap and Sanitary Chemicals* ran an ad for Associated Chemicals, one of the nation's leading producers of industrial chemicals, that proclaimed, "The End of the War Ushers in a New Era."[27] Indeed, the war created a different world, a world in which American interests abroad would expand greatly and in which chemical technologies would be seen as viable solutions to some of the most complex problems in the world, including disease and famine. The victory cast the United States as a major international force and affirmed the scientific principles Vannevar Bush espoused. The strategic funding of scientific innovation and the application of new technologies proved highly successful during the war and would serve as a catalyst for reviving the postwar economy at home and abroad. The chemical industry also embraced the postwar era with much enthusiasm. If anything, the war had demonstrated the importance of industrial chemicals, and the industry as a whole attempted to capitalize on its newfound power, which fundamentally changed entomological research and practices within the United States and beyond its borders.

But this world of possibilities and promise was fraught with danger. Beyond the borders of the United States lay two equally threatening things, insects and communists. Public health work during the war confirmed that insects could be vanquished with DDT. Public health officials and foreign policy experts were less sure if DDT was powerful enough to defeat the ominous threat of communism. What was certain was that the Cold War changed the meaning of insect control. In this new global struggle, DDT became the primary weapon against insects and communists.

Speaking before the National Health Assembly in 1948, Fred Soper called attention to the increasing links between national security, health, and international alliances. He stated, "The United States will need healthy nations as allies, able to produce and to fight for the common interest."[28] Those common interests, Soper suggested, were insect eradication, health, and freedom from communism.

These ideas were already being put into practice in Sardinia, a rocky island off the central west coast of Italy. Because of its Mediterranean location, Sardinia had been plagued with "*mal-aria*" — Italian for "bad air" — for hundreds, if not thousands, of years.[29] The main culprit of the disease was the *Anopheles labranchiae*, a native species of mosquito that bred in brackish waters along the coast and in stagnant ponds at higher elevations. Prior to the war, various attempts to reduce malaria rates on the island had been attempted, including a campaign started in 1925 by the Rockefeller Foundation's Lewis Hackett.[30] Remarkably, Hackett's campaign reduced malaria morality by 90 percent, but it failed to completely eliminate the disease or the mosquito vector. By the war's end, however, Soper's approach to species eradication dominated the public health landscape. The question was whether it could be done in Italy.

Founded in 1946, the Regional Organization for the Struggle Against the Anopheles in Sardinia (Ente Regionale per la Lotta Anti-Anofelica in Sardegna; ERLAAS) was a joint Italian and Rockefeller Foundation venture aimed at reducing the historic plague of malaria on the island. As the Naples campaign had, ERLAAS deployed DDT in massive quantities as a way to reverse the health impact of disease-bearing insects. More than anything else, the project's name symbolized the dramatic shift in how public health officials understood malaria. It was no longer a disease to be treated through therapeutic drug remedies or controlled by drainage. Rather it was a "struggle" against mosquitoes that sought to solve the problem of disease once and for all.[31] Similarly to Mussolini's *bonifica integrale*, the Sardinian project was more than a project against insects. It meant the promise of longer lives and economic growth. For the United States, ERLAAS served as an important outpost in the fight against communism.[32] If the Second World War can be thought of as a "public health war," then the Sardinian campaign was the first public health proxy war fought during the Cold War.

The Rockefeller Foundation utilized the lessons learned from its Naples experience, from its work in the South Pacific, and from Soper's regimented methods to attack the *Anopheles labranchiae* with military precision. The island of Sardinia, no bigger than the state of Vermont, was divided into eleven divisions, which were broken down into six sections, 57 districts, and 498 sectors. Using a suspended mixture of DDT and petroleum, teams of public health workers were sent across the island, armed with hand sprayers and bug bombs. Pilots dropped DDT from the air, spraying areas difficult to reach by car or foot. At the height of the project in 1949, ERLAAS employed 33,000 people, out of a total population of 1.2 million. By the end of the campaign the entire island had been sprayed with nearly 10,000 tons of DDT, reducing malaria rates dramatically: 75,447 cases in 1946 compared to 9 cases in 1951. Yet despite this achievement, ERLAAS failed to completely destroy the *labranchiae*. This minor technicality did not dissuade the public health community from celebrating its success.[33]

But the problem of malaria on Sardinia was very different from those previously tackled. The work Soper conducted in Brazil and Egypt, for example, focused on the eradication of nonnative species. The *Anopheles labranchiae*, however, was native to Sardinia, making the project of eradication a much more difficult task. Initially, this approach was not universally accepted. In fact, the first ERLAAS director, John Kerr, resigned because he did not believe that species eradication could be achieved on Sardinia. In a letter critical of the idea of eradication, Kerr wrote, "Call me a pessimist if you will, but the word impossible is in my vocabulary, and I intend to keep it there."[34] His replacement, John Logan, had no such reservations. Logan would go on to write the first large-scale history of the project, offering a glowing assessment of its accomplishments.[35]

Sardinians also began to question the massive spray campaign. Farmers complained that their sheep had been poisoned. Others claimed that aerial sprays led to massive fish kills or destroyed bee colonies. According to historian Marcus Hall, one Sardinian newspaper "spoke disparagingly of what it called 'American' methods . . . [and called] for more 'Italian' methods aimed at killing only enough mosquitoes to disrupt malaria transmission."[36] Spray teams encountered people resistant to DDT spraying, as had teams carrying out typhus and malaria experiments in Mexico during the war, and the encounters occasionally led to violent confrontations. For many, the ecological consequences and potential health risks of DDT far outweighed any tangible benefits it may have provided.

In the face of such opposition, IHD director George Strode wrote to Soper, claiming the Rockefeller Foundation "should stick to [its] guns."[37] And stick to its guns it did. Despite mounting concern about the eradication project, Rockefeller Foundation officials firmly believed in the transformative power of DDT. As a leading public health institution, the foundation was committed to DDT and investment in science and technology more broadly. That commitment carried enormous weight with many political and economic leaders around the world.

The Sardinian campaign was set against the continued euphoria over DDT and the war. For those involved in the development of DDT, the postwar years were filled with recognition and accolades. In 1946, Edward Knipling received the War Department's United States of America Typhus Commission Medal. A year later, he was awarded the King's Medal for Service in the Cause of Freedom by the British government. That same year, Fred Soper was named director of the Pan American Sanitary Organization (later renamed the Pan American Health Organization). In this position, Soper directed public health work throughout Central and South America until his retirement in 1959 and worked closely with the WHO malaria eradication campaign. And in 1948, Paul Müller, inventor of the miracle compound DDT, won the Nobel Prize in Physiology and Medicine for his contribution to public health. Despite warnings of its ecological hazards, DDT certainly had its champions in the highest levels of the scientific community.

To some, the quest to eradicate all mosquitoes paled in comparison to the economic development aspirations the public health work engendered. Once seen as an economic liability, ERLAAS sought to create a healthy terrain on which the free-enterprise system could take hold and flourish. "The island, formerly an economic liability," one Rockefeller Foundation official noted, was "emerging as Italy's new frontier. Sardinia [was] now a healthy place to live and work."[38]

Not surprisingly, the importance of this work was not lost on U.S. foreign policy officials. In fact, the ERLAAS project operated alongside and became enmeshed with the European recovery programs of the Marshall Plan. And for many, Sardinia was seen as "Italy's frontier zone," capable of alleviating population pressures in the north and laying the foundation for economic expansion, particularly within the agricultural sector. In 1949, James Zeller-bach, then head of the Economic Cooperation Administration (ECA) in Italy, told the American public, "The present population of one million [in Sardinia] could be doubled when malaria control is completed and extensive land improvements and irrigation projects finished."[39] Two years later, Leon Dayton, the head of the ECA's Italian mission, claimed, "Now that this disease and its causes have been eliminated by the fine efforts of the Rockefeller Foundation working through ERLAAS . . . the resources of the island may be used to enrich the Italian economy as a whole and the population increased by transferring people from Italy's overpopulated mainland."[40] "Developing Sardinia's resources," he argued, "might solve the whole Italian emigration problem."[41]

For Dayton, however, populating Sardinia and jumpstarting the Italian economy was part of a larger quest to integrate the economies of Western Europe and the United States. Speaking before the members of the Italian Chamber of Commerce for the Americas, Dayton proclaimed, "If we are to fulfill the industrial, agricultural and spiritual potentialities of Western Europe[,] integration is the way to that improved standard of living, and to that stability and peace to which the peoples of Western Europe are so justifiably entitled."[42] Of course, from Dayton's perspective, Italy would serve as an economic partner for the United States and as a barrier against communist influence.

The economic desires of American aid experts were not lost on Italian political commentators, particularly the editors of the Italian satirical newsweekly, *Candido*. Founded in 1945 by Giovanni Mosca and Giovannino Guareschi, and based in the economic center of the nation, Milan, *Candido* quickly emerged as an important voice critical of postwar Italian politics. Staunchly anticommunist, Mosca and Guareschi molded their journal to capture and expose the conflicting and completing agendas, both foreign and domestic, attempting to reshape the nation. Mosca and Guareschi were committed to exposing political hypocrisy on the right and left, making *Candido* a valuable independent voice in Italian political discourse. Other important writers joined *Candido*, including Marcello Marschesi and Carletto Manzoni. The use of simple line drawings and cartoons also provided the journal with a visual presence that set *Candido* off from other political journals.

In a common feature of the journal, Manzoni and Guareschi, writing under the pseudonym of Cesar and Spartacus respectfully, debated Italian politics from different political perspectives. Manzoni's Cesar wrote about politics from "the right," while Guareschi's Spartacus wrote from "the left." The U.S. presence in Italy certainly captured their attention.

Manzoni and Guareschi set their sights on Leon Dayton's visit to Milan in the October 29, 1950, edition of *Candido*. Manzoni praised Dayton's efforts to encourage Italian industrialists to use ECA funds, which had "eliminat[ed] all the obstacles of production." These obstacles included the presence of the communist party. Manzoni argued that "Dayton want[ed] to improve the life style of the workers with higher weekly salaries." But Manzoni was critical of U.S. and Italian industrialists' attempt to buy Italians and of Dayton's authority. He wrote, "Dayton determined the funding, not the Italians." Guareschi was far more critical of Dayton's presence in Italy, accusing him of replacing the church and the monarchy. Guareschi proclaimed, the "Cardinal kisses Dayton on the ring, not the other way around."[43]

Another article by Guareschi from the same issue was equally critical. He believed Dayton was trying to "Americanize" Italy, calling him a "colonizer" and characterizing his policies as "anti-Italy." Guareschi, however, was not critical of the U.S. presence in Italy. Rather it was Dayton's exercise of power within Italy that shaped Guareschi's thinking about the ECA. "Please Mr. Dayton," Guareschi implored, "go back home. No one likes you from the left and the right."[44]

Guareschi's anticommunism captured the attention of a number of influential Americans, including Henry Luce. In 1952, *Life* published a lengthy story on Guareschi, calling him the "Anti-Communist Funnyman." "When . . . Guareschi gets on the subject of Communism," *Life* reported, "his eyes bulge, he flails his chest and he lapses into the dialect of his native Parma." As editor of *Candido*, *Life* informed its readers, Guareschi has been "subjecting Italy's Communists to such a withering barrage of ridicule that the captions on his cartoons have become conventional catch phrases in nearly every café in the country."[45]

Guareschi was not the only cartoonist on the staff. His depiction of a com-

"L'Industria Italiana, I Parassiti e . . . il D. DayTon" (Italian Industry, Parasites, and DDT),
drawing by Carletto Manzoni, from *Candido*, November 12, 1950.

munist as a three-nostrilled monster, however, is perhaps the most recogniz-
able image from the journal. Other artists and writers contributed to *Candido*'s visual legacy, including Carletto Manzoni, who in one image connected
anticommunism, Italian redevelopment, and DDT.

In November 1950, as mosquito eradication in Italy was reaching its apex
and debates about economic development and communism continued to
dominate the Italian political landscape, *Candido* published an illustration
depicting the transformative promise of DDT. Framed by a caption that reads,
"Italian Industry, the Parasites, and DDT," the image, which is remarkably
similar to Brandt's photographs taken six year earlier, shows a medical officer

dressed in a white lab coat spraying DDT on the Italian nation depicted as a woman dressed in rags. The nation towers over the health official and looks somewhat displeased at his efforts. Yet in this case, DDT was not used to rid the nation of lice or mosquitoes, but communists. With each pump of the spray gun, the Soviet-style hammer and sickle, which decorates the nation's garment, falls to the floor, freeing Italian industry and the nation from the ravages of communism. Presumably, once the nation had freed itself from Soviet control, Italian industrialists could enjoy the fruits of the free market.[46]

The image of freedom brought by the spray gun perfectly captured a particular moment in the history of DDT. Only a few years later, many Americans would be clamoring for freedom *from* the spray gun. Yet in both cases, Cold War politics, economic development, and conflicting ideas about nature would inform views on the risks and benefits of DDT. Still, by the early 1950s the impact of the Sardinian campaign was being felt far and wide.

As the results of the ERLAAS campaign began to spread, the project's administrator, John Logan, received a letter from the recent recipient of the Nobel Peace Prize, Lord John Boyd-Orr, who received the prize for his work on food nutrition and world famine and served as the first head of the U.N.'s Food and Agricultural Organization from 1946 to 1949. Boyd-Orr offered his congratulations to Logan on the "magnificent job done in the face of almost insuperable difficulties in eliminating malaria." But rather than looking back at the success of the ERLAAS campaign, Boyd-Orr wrote, "The elimination of malaria should be considered as merely the first step to a grand program of bonifica." Sardinia, he suggested, was important because it was "somewhat typical of a large area of the earth with more than half of the population of the world," meaning that Sardinia was important not because it was an isolated island, but because it was representative of much of the world: Sardinia was poor, undeveloped, and diseased. For Boyd-Orr, the ERLAAS project in Sardinia offered the possibility and promise of a better world.

"The first step," Boyd-Orr continued, "has already been taken. . . . The next step is to develop the natural resources to increase the wealth and the standard of living of the people." This mission, however, was not just to improve the lives of Sardinians, although that was part of the equation. The mission was a "challenge to western civilization" to use "modern sciences" that were "capable of eliminating . . . diseases and creating wealth." Boyd-Orr asserted, "The future of western civilization depends upon its acceptance of this challenge and its willingness to carry through a global bonifica to bring about social contentment by realizing the just hopes and aspirations of these people. The carrying through of such global schemes will benefit the highly industrialized countries as well as the undeveloped countries because vast quantities of industrial products will be needed to bring about a rapid increase in this real wealth of the world. This will bring about rapidly expanding world economy needed to absorb the products of modern industry and agriculture."[47]

There is no doubt that for people like Boyd-Orr, John Logan, or Fred Soper, along with a host of other participants, observers, and Sardinians, the eradication campaign was an overwhelming success. The rapid decline of malaria in Sardinia was not only remarkable but transformative. Sardinia was the model on which public health officials developed global malaria control programs. But more significantly, the aspirations for physical and economic health espoused by public health officials, national governments, and transnational organizations like the Rockefeller Foundation and the United Nations created unrealistic expectations for the future use of DDT and the model of insect control "perfected" by ERLAAS.

The fruits of public health and redevelopment projects in Italy were on full display in the economic capital of the United States, New York. On September 10, 1951, the nation's largest department store opened its doors to a unique event called "Italy-in-Macy's." Held over a six-week period in September and October, Italy-in-Macy's showcased some of the finest hand-crafted Italian goods for the American buying public. Driven by economic cooperation between Italian producers and American consumers, the Italy-in-Macy's event was also deeply political. In a letter to Leon Dayton, chief of the Marshall Plan's Italy mission, Macy's president, Robert Weil, claimed the exposition of Italian goods would "be a vivid and realistic application, by American and Italian private enterprise, of efforts 'beyond the Marshall Plan' to strengthen the economy of the free world." Weil proclaimed that the Italy-in-Macy's campaign would "re-confirm the virility of the unique Italian skills, handed down from father to son." He maintained, "It will be striking proof of Italy's very real 'Second Renaissance.'"[48]

The exhibit contained a full-size Venetian gondola, a reproduction of St. Peter's Church, and a model of Christopher Columbus's ship. Along with these visual displays, Macy's employed Italian craftsmen to demonstrate glass-blowing techniques, leather craft, and ceramics-making. A donkey cart representing "old" Italy was adorned with images of President Truman and George Marshall, reminding attendees of the "new" political and economic connections between the two nations. On the first day of the event, nearly 25,000 people attended the festivities. "These displays of Italian culture," the New York Times reported, "are encouraging and significant in these times of crisis headlines. . . . The Macy's exposition is not only a gratifying example of the effectiveness of Marshall Plan aid; in another sense it reveals a unique historical trait of the American character: the willingness to offer to former foes the hand of friendship in working toward the common goal of peace in the world."[49]

Italian economic health seemed to confirm the logic of eradication and the transformative power of DDT. Although Marshall Plan funds bankrolled much of the recovery program, public health programs similarly contributed to Italian economic health. Undoubtedly, the Sardinian campaign generated enormous interest, but DDT was also used throughout Italy to vanquish disease-bearing mosquitoes. Massive spraying programs in the Po River Valley and the Pontine Marshes reduced malaria rates considerably, thus creating a healthy population of workers throughout Italy. If the Macy's-in-Italy event had an underlying message, it was that public health and economic development went hand-in-hand.

Once malaria had been brought under control, however, no one seemed to ask if the effort had been worth it. Although critics of the Sardinian campaign, like Kerr, questioned the concept of species eradication, very few asked if using DDT in such massive quantities was worth the risk of coating entire regions, or an entire island in the case of Sardinia, with an ecologically dangerous chemical. Certainly some Sardinians expressed concern about the massive spray campaign. But the success of the campaign clearly overwhelmed the voices of dissent.

Despite the popular enthusiasm for the chemical, certain ERLAAS scientists were well aware of the scientific literature that questioned DDT's ecological consequences and long-term impact on human health.[50] Moreover, in their study released in 1946, the same year ERLAAS began, Paul Putnam and Lewis Hackett detailed their malaria control efforts in Sardinia prior to the war. Their report indicated that improved medical diagnoses, drug treatments, water drainage, and the strategic deployment of Paris green reduced malaria mortality by 90 percent.[51] While mosquitoes survived, the transmission of the malaria parasite had been disrupted, bringing the death toll down to historic lows, a number that continued to decline with eradication.

But the postwar Sardinian campaign took place in the context of an increasingly vehement struggle against communism. In waging the war against communism, President Truman strengthened U.S. military commitments abroad. The acknowledged successes of the Rockefeller Foundation's public health work made humanitarian and economic aid an important part of U.S. Cold War foreign policy. In 1950, hoping to emulate the successes of the Marshall Plan and the work of the Rockefeller Foundation, President Truman launched the Technical Cooperation Administration, which would administer a project known as the Point Four Program. In outlining the program, Truman acknowledged the importance of malaria control in developing strong economies. Speaking before an audience at the American Newspaper Guild, Truman claimed that DDT had rid the Terai District of India of mosquitoes. He announced that "families who were refugees from malaria only a year ago are back in their homes, and their fields are green again."[52] Indeed, the links between agricultural development and health were deeply imbedded in Point Four, a program that Truman believed would be a "a revolution against hunger, disease, and human misery."[53] And, as we shall see, Truman looked to the Rockefeller Foundation for assistance, not simply because of its commitment to public health, but also because of its innovative work in the field of agricultural development.

Under the Eisenhower, Kennedy, and Johnson administrations, foreign aid would be one of the more critical aspects of U.S. Cold War policy. Because of success in eradicating mosquitoes and disease there, Sardinia was increasingly seen as the launching point for a global revolution, one that would shift U.S. Cold War interests from Europe to Asia, Latin America, and Africa. The Sardinian campaign became the public health model that would support American efforts to "develop" many parts of the world. This shift would ensure that DDT was on the front lines, defending the nation's interests abroad, all in the name of humanitarian aid.

News of the Sardinian campaign filtered into the American consciousness through selected publications and films. Not as overtly dramatic or widely publicized as the Naples campaign, reports on Sardinia nevertheless emphasized similar humanitarian themes, albeit within a vastly different political climate. Cold War tensions reshaped the parameters of U.S. foreign policy, sharpening the distinctions between the free and unfree, the healthy and the sick, and the developed and undeveloped. Many press reports about Sardinia captured these sentiments as a way to call attention to the project's success, while underscoring the need for long-term development projects.

Initial reports seemed promising. In 1951, former *Life* editor Paul Deutschman wrote a story for *Harper's* magazine about his travels to Sardinia and the island's resurgence, brought about by ERLAAS. "The purification of the island," he wrote, "has made all of it habitable and has opened its natural treasures to economic use. . . . That American money, techniques, and drive were the instrument of this 'miracle' (even a Sardinian Communist will take you aside and call it that) is no surprise." Deutschman was on his second trip to the island (his first was as a serviceman during the war), and for him, the transformation was nothing short of astonishing. "The funereal grayness of wartime was gone," he wrote. "There was a rush of civilian color along the inner street." The public markets closed because of the war were "wide open and humming."[54] By "purifying" the island and the scourge of malaria, ERLAAS released the economic energies of Sardinians, creating a healthy environment in which market capitalism could develop and prosper.

But was this true? Was change simply that easy? Despite the arduous work of ERLAAS officials and sprayers, could the economic dormancy of Sardinia simply be reversed in a few short years through eradication?[55] For some, the Sardinia experience offered hope and promise. For others, it was a warning that eradication itself could not solve complex economic problems.

To avoid any ambiguity about the transformation of Sardinia, the U.S. government funded two films, *Adventure in Sardinia* (1950) and *Isle of Hope* (1953).[56] Produced by the ECA, *Adventure in Sardinia* told the story of how malaria had suffocated the economic potential of the island. Once freed of the disease, Sardinians could enjoy the economic benefits of a market economy. *Isle of Hope* had, perhaps, a much more explicit political message. Produced by the Mutual Security Agency, *Isle of Hope* told the story of a boy who rejects the sheep-herding tradition of his family in order to learn about the new technologies introduced by Marshall Plan funds. He is enthralled by the increased coal production of his country, which is driving postwar economic expansion. Eventually, he returns to farming, but he brings with him the knowledge of modern agricultural techniques that allow him to grow crops more efficiently and with greater yields.[57] These films reinforced the connection between physical health and economic and political freedom, creating linear narratives of progress and development that came to define the politics and practices of American cold warriors.

But Italian newspapers questioned this development model, suggesting that health and foreign aid did not produce the desired political outcomes. In 1953, *Esteri* opined, "American public opinion finds it even more difficult to understand why, after so many years of aid and so many millions [of] dollars spent, Communism has not yet disappeared from Italy." An editorial in Rome's daily newspaper, *Il Messaggero*, proclaimed, "The Marshall Plan did not and could not propose to change Europe's face, to cancel in only two or three years the results of events that have taken place in the Old Continent during several dozens of years." And while the editorialist conceded, "Communism has not yet been wiped out in Italy," he claimed that the "way to rehabilitate our social and political structure cannot be picked at random."[58]

In other words, the United States could not legitimately impose a political structure on Italy. *Esteri* proclaimed, "It would have been naive to expect that with aid alone we could restore entirely Italy's prewar wealth."[59]

Henry Luce continued to concentrate on the Italian situation; it was never far from his mind. In 1953, the same year his wife, Clare Boothe Luce, became the U.S. ambassador to Italy, *Fortune* published the commissioned work of Costantino Nivola, a Sardinian-born artist who chronicled the story of Sardinia's historic transformation in a series of pen and ink drawings. The images—twenty-one in all—and associated text carried the reader through a timeline of the island's history, from its historic battles with the Romans to the outcome of the malaria campaign, which, *Fortune* reported, "has been the greatest event in their history."[60] Nivola's illustrations depicted a pastoral landscape of the Sardinian countryside, the stony mountain villages, and vestiges of a sheep-herding culture. He painted images of women dressed in long black dresses, mourning the deaths of their sons and husbands. "Death, by violence or diseases," one caption reads, "is familiar to all Sardinian women." Images of men searching for mosquito breeding grounds or spraying DDT follow. The text informs readers that "no building—house, store, church, prison, shed—was left unsprayed; in all 337,000 separate buildings on the island were treated inside and out with DDT."

Toward the end of the piece, readers were confronted with two disparate images, one of the bustling port of Cagliari, the island's largest city, and the other of a group of men sitting in a piazza. The image of Cagliari depicts a healthy and vibrant economy, like the one envisioned by Soper, Logan, and ECA chief Leon Dayton. The image was the realization of a successful public health campaign, where movement, rather than stasis, expressed the realization of a modern economy. Yet the accompanying caption suggested all was not well. "The processes of politics and society move at a slower, less manageable pace than the technical campaign against the disease," *Fortune* suggested.

The second image portends something much more troubling: unemployment, communists, and insect resistance. The men sitting in the piazza were not collectively enjoying the fruits of their labor; they were, in fact, unemployed, presumably losing their jobs once ERLAAS disbanded in 1951. Not only had the unemployed returned to the piazza, so had the "flies," which, as the caption indicates, had developed a resistance to DDT as early as 1947, the result of an evolutionary process that had become widely known by the end of the ERLAAS campaign.[61]

The image of the return of flies and unemployment to Sardinia perfectly

The promise of eradication in Sardinia. The spray campaign would free Sardinia from malaria and economic sluggishness to create a vibrant, bustling economic center. Image of Cagliari by Costantino Nivola, originally published in *Fortune*, March 1953. Courtesy of the Nivola family.

captured the structural inadequacies of the ERLAAS strategy. In their analysis of the Sardinian campaign, *Fortune* editors called into question DDT's capacity to jumpstart a sagging economy and the long-term consequences of the chemical as an effective killer of insects. And while DDT killed mosquitoes and saved lives, the technological and chemical approach to insect eradication undermined alternative approaches to malaria control and economic development. *Fortune* observed, "The over-all Italian program for rehabilitation and reclamation in southern Italy . . . has been slow to materialize." Alongside this economic problem, *Fortune* heard the "communists drum their steady warning of U.S. imperialism."

Yet *Fortune*'s story was not meant to dissuade future investment in Italy. On the contrary, the story was intended to encourage investment. Henry Luce's vision of the American Century required American business leaders to take an active role in world affairs. Rather than sit idly by, Luce demanded that American businesses engage in a global economy, although a global economy

Failed expectations in Sardinia. This image illustrates the limits of the eradication campaign. Not only was the Rockefeller Foundation unsuccessful in eradicating the *Anopheles labranchiae*, but the campaign did not live up to the expected promise of economic development. Image by Costantino Nivola of men sitting in the piazza. Originally published in *Fortune*, March 1953. Courtesy of the Nivola family.

shaped by U.S. self-interest and economic power. So instead of being seen as a failure of U.S. technical aid, the "DDT in Sardinia" piece served as a warning that health by itself could not reverse economic underdevelopment. Only long-term engagement with Sardinia, or Italy, for that matter, could ensure that American economic and political interests would be served. Disease prevention was only the first step to a much longer process of engagement.

REPRESENTATIONS OF DDT'S OVERSEAS use expressed a series of ideas that reinforced assertions about the scope and limitations of U.S. power in the world. Initial depictions of the typhus campaign in Naples carried with them visual cues that promoted the humanitarian aspects of the war while at the same time bolstering the political imaginings of U.S. involvement in postwar Europe. Utilizing a newfound weapon against insects became a way to define U.S. interests abroad at a time when Cold War contests over political allegiances intensified. The Sardinian campaign was not only an attempt to modernize the Italian island, but an effort to demonstrate the modernizing capabilities of the United States. Saving lives through eradication was a way to confirm U.S. beliefs about modernity, technology, and the strength of America abroad. DDT's mosquito-killing proficiency only added substance to the belief in U.S. foreign aid, international development, and technology as effective instruments of change.

In Naples and Sardinia, the war against insects and disease demonstrated the power of the new insecticide. DDT killed insects and saved lives. Although the two projects differed significantly in their approach and the means of delivering DDT, each drew attention to the viability and usefulness of DDT for public health. Equally significant was the role the Rockefeller Foundation played in testing and adopting DDT as the main weapon against insects and disease. Relying on an extensive and strictly regimented spray strategy, the Sardinian project bolstered the claims made by ERLAAS officials and other malaria specialists that DDT could profoundly change the health and economic future of the island. Thus, in a few short years, the Rockefeller Foundation's experience in Italy heightened the expectations about DDT and insect eradication on a global scale.

Not to be outdone, the foundation's agricultural team, operating out of a small research station outside Mexico City, also found novel ways to incorporate DDT in a transformative project that fundamentally changed the way many people around the world grew food. Just as the quest to eradicate disease had, the Green Revolution would symbolize the triumph of human technology and represent the excesses of technological modernity. Often seen as parallel development schemes, postwar public health projects and the Green Revolution intersected in interesting and unforeseen ways. Not surprisingly, DDT lay at the center of these collisions.

3

SCIENCE IN THE
SERVICE OF AGRICULTURE
DDT and the Beginning of the
Green Revolution in Mexico

*[T]o respond nobly and patriotically . . . for
producing an abundance of foodstuffs for the nation.*
—The League of Agricultural Communities of Jalisco and
the Federation of Small Property Holders, Jalisco, Mexico, 1953[1]

Following the election of Mexican president Manuel Ávila Camacho in 1940, U.S. vice president–elect Henry A. Wallace traveled to Mexico City to attend the inaugural festivities. As Wallace was the first high-ranking member of the U.S. government to attend a presidential inauguration in Mexico, his trip symbolized President Roosevelt's renewed commitment to building better relations with Mexico (and Latin America), better known as the Good Neighbor Policy. Roosevelt believed that the United States, through engaged diplomacy and economic cooperation, could help fulfill the nationalist and social democratic promises of the Mexican revolution, while at the same time promoting industrial development with American capital.[2] Reporting on the trip, *Life* asserted, "The Wallace excursion to Mexico is in part a belated attempt to make up for the many decades during which the U.S. alternated between indifference and exploitation."[3]

Wallace's visit also set in motion a process that dramatically transformed how humans grew food. After the inauguration, Wallace spent the last weeks of 1940 traveling through the Mexican countryside surveying the agricultural landscape. Having come from an Iowan farm family, Wallace had distinguished agricultural credentials. He served as the editor of his father's weekly agricultural journal, *Wallace's Farmer*; founded Hi-Bred Corn, producers of industrial hybrid corn seeds; and served as the U.S. secretary of agriculture, as his father had. Believing in the promise of agricultural science and technology, Wallace was troubled by what he found in making an assessment of the Mexican countryside. "The atmosphere of rural Mexico is amazingly like Bible times," Wallace wrote. Corn was processed "just as it was two thousand years ago in Palestine." Ever the progressive farmer, Wallace went on: "From the standpoint of the great bulk of people who live on the farms in Mexico, I can't help thinking that efficient corn growing is more important than almost anything else in Mexico."[4]

The inefficiencies of Mexican food production concerned U.S. policymakers. Writing to Secretary of State Cordell Hull, George Messersmith, the U.S. ambassador to Mexico, warned, "The situation with respect to corn here is serious, and when I mean serious, I mean very serious."[5] During Ávila Camacho's first year in office, Mexico suffered from some of the worst harvests on record. Given the tenuous but improving relations with Mexico, Roosevelt's State Department continued to monitor the food situation. The State Department also believed that American aid could be best served by a nongovernment agency, particularly one that had existing relations with the Mexican government, rather than through direct federal intervention.

Upon his return to Washington, D.C., Wallace met with Rockefeller Foundation president Raymond Fosdick to discuss the agricultural situation in Mexico. Wallace hoped that the foundation would undertake a new project to help alleviate the Mexican agricultural crisis. Fosdick remembered Wallace's saying, "Health by itself is not enough. With no improvement in agriculture, as the death rate declines, the inevitable result is a lowered standard of living."[6] The seeds of a global agricultural revolution were sown in this pivotal moment.

Analogous to the public health programs of postwar Europe, food production became another measure of defining health and modernity in the postwar period. But unlike the targeted effort to eradicate mosquitoes with DDT, the Mexican experiment widened the terrain on which the technological and economic models of development would be deployed. From the Rockefeller Foundation's point of view, increasing agricultural production raised a host of issues, including the immediate challenge of how to create an agricultural program from the ground up. But perhaps more importantly, the foray into agricultural work required the foundation to confront a myriad of scientific, technical, environmental, and political issues that complicated its program. Although DDT was just part of a larger agricultural revolution, exploring the chemical's role in the early formation of the "development project" helps us better understand how DDT became seen as a vital component of U.S. foreign policy in the postwar years. The overlapping and, at times, conflicting sciences of public health and agricultural development were committed to improving human health on a global scale.

For Mexico's political leaders and economic elites, agricultural development was a necessary step in transforming the nation's economy.[7] Upon assuming office, Ávila Camacho called on the Mexican agricultural sector to be the "foundation of industrial greatness."[8] He, along with other Mexican leaders, believed that healthy, well-nourished citizens were a critical element in transforming the economic fortunes of the nation.[9] The pursuit of increased outputs led the Mexican administration to embark on an agricultural system that required sizable technological inputs and "scientific knowledge" to generate the amount of food deemed necessary for an industrializing nation.

By turning to the Rockefeller Foundation, Mexican leaders sought to reverse the inefficiencies of the past through modern agricultural science and technology. Hybrid seeds, tractors, fertilizers, and chemical pesticides provided the means to awaken a dormant agricultural landscape. They also served the interests of Mexico's political and economic elite, who often had greater access to capital and owned better agricultural lands.[10] The use of these new technologies marked the maturation of a modern nation-state. Changes in Mexican leadership and state policy—reflecting the new belief in science and technology as arbiters of the modern world, the narrowing of entomological knowledge around synthetic chemicals, the expansion of the postwar chemical industry, and, finally, the foreign policy aspirations of the U.S. government—all contributed to the radical transformation of Mexico's rural landscape. The paradox, of course, is that the agricultural system embraced by Ávila Camacho and sponsored by the Rockefeller Foundation would put the bodies of many Mexicans at risk, creating what Angus Wright has called the "modern agricultural dilemma."[11]

BEGINNING IN 1942, the Rockefeller Foundation launched an investigation of the Mexican agricultural landscape by sending three agricultural specialists on a six-month fact-finding mission. Richard Bradfield, from Cornell University, Elvin C. Stakman, a plant pathologist from the University of Minnesota, and Paul C. Mangelsdorf, from Harvard University, surveyed Mexico's agricultural lands and institutions. All three had impeccable scientific credentials, yet it was Stakman who, over the next few years, would influence the direction of agricultural development in Mexico and the United States. A committed scholar and teacher, Stakman encouraged one of his students, Norman Borlaug, to embark on a career on plant diseases. Stakman's instincts for talent were right, as Borlaug's research and advocacy for modern agriculture would radically transform rural lands throughout the world. Stakman also became one of the most outspoken advocates and defenders of the Rockefeller Foundation's agricultural work.[12] And, as president of the American Association for the Advancement of Science, Stakman was a tireless promoter of American scientific innovation to alleviate the problems of the world.

Fieldwork by what was officially called the Survey Commission on Mexican Agriculture took the three agricultural advisers to sixteen of the thirty-five Mexican states. Often traveling by truck, horseback, or, at times, by foot, they met with Mexican agricultural leaders, as well as local farmers, in order to grasp the complexity of food production in Mexico. Upon their return, they claimed that "the most acute and immediate problems . . . [are] the introduction, selecting, or breeding of better-adapted, higher-yielding and higher quality crop varieties [and a] more rational and effective control of plant diseases and insect pests."[13] The advisers' work set the parameters for a new Rockefeller Foundation project, the Mexican Agricultural Program (MAP).

While MAP was a new undertaking for the foundation, the organization had vast experience working in the country. Beginning in 1927, the Rockefeller Foundation developed and administered a number of public health projects in Mexico to reduce hookworm, typhus, and malaria. Although considered successful in the sense that they reduced disease, these projects were not without their problems, as both Mexican and Rockefeller officials were at times exasperated by their working relationship, leading one historian to call the partnership a "marriage of convenience."[14]

The postwar years also brought tension as the Rockefeller Foundation's International Health Division (IHD) began a mosquito eradication campaign. Wilbur Downs, the IHD director in Mexico, recalled that after spraying the town of Mixquic with DDT, foundation officials were "literally the town heroes."[15] But according to historian Marcos Cueto, local "opposition to spraying had been strong, and sometimes armed."[16] Not only were these differences fought out on the ground, but conflicts also arose within the bureaucratic hierarchy of the eradication program. Officials from the IHD often had ideas about disease prevention and insect eradication that conflicted with those of local or national health officials within Mexico. Despite the foundation's leadership role, its public health programs were not without some contentious moments.

To many in the Rockefeller Foundation, the fact became obvious that public health and agricultural programs in Mexico were deeply intertwined. Stakman recalled that "medical people . . . [began] to realize that better nutrition was essential if they were going to improve the health [of Mexicans]. . . . It was perfectly evident that many of the diseases that the people had in Mexico were due partly to the predisposing effects of poor nutrition or of hunger."[17] Connecting these projects further was the relationship between humans and insects. Indeed, one of the major strategies for improvements in public health and food security was insect control. In other words, insects, rather than disease or agricultural inefficiencies, challenged the Rockefeller Foundation in its quest to improve human health, whether in Italy or Mexico. But public health and agricultural experts had different ideas about how to deal with the insect problem. On Sardinia, the foundation attempted to eradicate mosquitoes with a massive DDT spray campaign. In Mexico, DDT was not the central chemical actor, but just one of an assortment of synthetic chemicals used to control insect pests. In this regard, DDT was simply part of a larger agricultural revolution that engendered a global shift in how "modern" societies grew basic grains. But because DDT was part of a larger technological and political project, the significance of its use for agricultural development cannot be marginalized.

Despite the Rockefeller Foundation's modest financial commitment to the project, MAP officials sought a complete transformation of Mexico's agricultural landscape.[18] "The objective is not unlike those of other foundation projects dealing with public health improvement," the foundation reported. "Indeed, the purpose may be regarded as a parallel one, since nutrition is closely related to health. But it is not only adequate nutrition and good health that depend upon the yield of Mexico's farms. The whole Mexican economy rests on this output, and no enterprise is more basic to the national welfare than the improvement of agriculture."[19] Similarly to the desired outcomes of the Sardinian eradication campaign, the objectives of MAP were not simply to increase crop production. The main objective was to create a thriving economy in line with U.S. interests. For that reason, DDT became immersed in an agricultural system designed to integrate U.S. interests abroad. Increasing crop yields was simply the first step in a much larger political and economic revolution.

In 1943, J. George Harrar, a plant pathologist from Washington State University, was named MAP's first director. (He would later be named the Rockefeller Foundation's president in 1961.) Operating under a semiautonomous agency, the Office of Special Studies, within the Mexican Department of Agriculture, the Rockefeller Foundation provided funds for MAP staff and agricultural materials. The Mexican Department of Agriculture contributed to the program as well, furnishing buildings, land, and support staff, including students and laborers. Much of the research was done at the National School of Agriculture in Chapingo, approximately twenty-five miles northeast of Mexico City, although experimental stations would be located throughout the country. Harrar then hired a team of U.S.-trained scientists, including geneticist Edwin J. Wellhausen, soil scientist William Colwell, entomologist John J. McKelvey, and, in 1944, plant pathologist Norman Borlaug.

Having been educated in the land-grant system, a system deeply engrained in a specific technical approach to agricultural production, members of the Rockefeller Foundation's agricultural project brought with them a set of assumptions about the superiority of American science and the backwardness of Mexican agriculture.[20] Stakman expressed this sentiment more directly. He recalled, "There had been virtually no research scientists in Mexico before the beginning of the campaign." "The very fact that [Mexico] asked for help from outside of the country," he later claimed, "was a recognition of the very obvious fact that they didn't have at that time the competence within Mexico to do what obviously needed to be done."[21] Harrar acknowledged, "There is very little modern agriculture practiced in Mexico," citing "adherence to tradition; inadequate educational facilities; lack of finance; primitive machinery and lack of research facilities" as the main reasons for Mexico's agricultural deficiencies.[22] For Harrar and the MAP team, these shortcomings were in stark contrast to the agricultural bounty of wartime America, a period one historian called the "golden age" of U.S. agriculture.[23] Indeed, the tremendous crop yields generated during the war years encouraged agricultural scientists to believe that their highly mechanized approach to crop production was vastly superior to alternative models, even though they were less than a decade removed from the ecological catastrophe of the Dust Bowl.[24]

Eager to begin work, Harrar and his team arrived in Mexico in the summer of 1943. Much of the work focused on improving the yields of the three major Mexican crops, corn, wheat, and beans. After an initial survey, the Rockefeller Foundation believed that seeds collected from Mexican sources could not generate the yields necessary for large-scale development. As a result, the foundation embarked on a program to find and use "the best seeds" available, obtaining seed stock from such places as the United States, Australia, Kenya, Africa, and Brazil. In 1944, Norman Borlaug began experimenting with these stocks, ultimately producing hybrid wheat that would revolutionize grain production for the next forty years.[25]

Entomological research and insect control became a focal point of the Mexican program as well. Harrar noted, "I think we might make great contributions to Mexican agriculture by attacking important insect problems and some of these probably could be made in a relatively short time."[26] Led by John McKelvey, a number of entomological studies were developed by the foundation to determine the effectiveness of pesticides against specific insect pests. In conducting these experiments, researchers had to deal with innumerable variables, including climate, geography, seasonality, aridity, humidity, rainfall, insect ecology, plant disease, and pesticide application. These early studies attempted to catalogue the complexity of a "foreign" agroecosystem with the aspiration that American-made pesticides could quickly control that agroecosystem.[27] From the foundation's perspective, control meant reversing the "economic losses" of insect infestation, thus realizing the full economic potential of an agricultural field for the farmer and nation alike. By connecting insect control with economic prosperity, the Rockefeller Foundation worked "to demonstrate the economic value of fungicides and insecticides in crop production and to encourage farmers to consider these activities as standard agricultural practices for certain specific crops."[28]

This entomological research revealed a troubling situation. "Insect pest problems," the foundation reported, were "numerous and serious," and resulted in a decline in agricultural yields and economic loses.[29] McKelvey's team identified insect pests for the three main staple crops, including grubs, thrips, leafhoppers, ear worms, red spiders, aphids, Apion pod weevils, and Mexican bean beetles.[30] MAP officials worked within complex ecosystems in a way that most public health experts had not. Anopheles mosquitoes were one thing; the myriad insects affecting agriculture were quite another. And it was clear that risks and outcomes of insect control differed significantly. Mosquitoes had the potential to kill humans; thrips, leafhoppers, and bean beetles, on the other hand, could destroy crops. This does not suggest, however, that insect control for agricultural production was less risky—it was not. But it operated on a difficult calculus.

Along with field research, the Rockefeller Foundation constructed an insectary to further the "study of life histories of some of the lesser known insect pests." Examining the physiology, breeding patterns, and life cycles of insects provided entomologists with much-needed information about the threat insect pests posed to crop production, a threat that was "of major importance to agriculture."[31] Once these data were gathered, the question facing

MAP entomologists was how to control insect pests. Should entomologists encourage biological controls, or, given the transformative nature of MAP, would chemical controls be more acceptable? If the latter, then how should dangerous chemicals be introduced into what was perceived as a backward agricultural system?

Within the United States, many agricultural experts advocated the use of chemical controls. Paris green, pyrethrum, and lead arsenate were among the most common pesticides used. The war expanded this chemical terrain. With the successful use of DDT, the new synthetic pesticide would soon be used for crop protection, even though little was known about the chemical's ecological hazards.

Operating under the assumption that modern farming required modern chemicals, the Rockefeller Foundation readily adopted synthetic pesticides as a central part of its agricultural development project. While the Second World War limited the availability of certain pesticides, by 1946 the foundation reported, "An effort is being made to control insects both by chemical and sanitary measures. In this connection the newer insecticides are being secured and evaluated for use in Mexico."[32] Another report claimed, "Extensive tests with a number of compounds are in progress." And by 1949, the foundation indicated, "A significant portion of the project is devoted to experiments with insecticides and the development of dust and spray techniques for use both in the field and in granaries. When the insect pests of a specific crop are known, control experiments are made in order to establish treatment schedules."[33]

Other entomologists, however, were less committed to chemical controls. In the United States, for example, a number of entomologists remained convinced that in the long run, biological control of insect pests was the most effective means to protect crops. However, these voices were in the minority. Agricultural officials within the USDA championed the use of agricultural chemicals over methods of biological control.[34]

Within Mexico, the situation was much more uncertain, since chemical controls were something of an anomaly, rather than a norm. In 1944, the Mexican secretary of education published a book by U.S.-born ornithologist and one-time head of the Pan American Union William Vogt. *El Hombre y la Tierra (Men and the Land)* would become an influential text among more-conservationist-minded Mexican leaders.[35] Although much of *El Hombre y la Tierra* dealt with soil conservation and how to improve soil fertility, Vogt identified one of the leading causes of agricultural losses: insect pests. "The worst enemies of men," he wrote, "are insects." But rather than promote the use of chemicals to control insect pests, Vogt suggested that birds could do the work of pest control far more effectively and at far less cost than human technologies. "They work for you," Vogt told his readers. "They eat insects—millions and millions of insects—that attack food." Poor harvests, Vogt reasoned, were caused by the destruction of bird populations, not by insects, suggesting that the Mexican farmer "kill[ed] his allies" because he believed birds were responsible for crop destruction. Vogt argued that the protection of birds was "one of the most important things that [could] be done to enlarge the wealth of Mexicans."[36] Despite promoting the same economic outcomes that the Rockefeller Foundation and the Mexican administration had, his criticism of industrial agricultural largely fell on deaf ears.

In the postwar years, Vogt emerged as an important voice critical of human dominance over the natural world. His book *The Road to Survival* (1948) created a stir because of his critique of modern agriculture and concern over population growth. "Every year," he wrote, the world is "producing less food," while human populations continue to rise at an alarming rate. Citing examples from the depletion of soil nutrients to the "lost song" of extinct birds, Vogt questioned the damage human societies wrought by industrial capitalism and called for humans to live in "ecological balance" with the natural world.[37] He was fearful that population growth would soon outpace food production, giving rise to starvation and food shortages on a massive scale. In fact, Vogt's neo-Malthusian arguments compelled Rockefeller Foundation president Chester I. Barnard to question the foundation's agricultural work in Mexico.[38] This was not enough, however, to suspend the Mexican project.

Fairfield Osborne, president of the New York Zoological Society, made a comparable critique in his book, *Our Plundered Planet* (1948). "Technologists," he wrote, "may outdo themselves in the creation of artificial substitutes for natural substances." Polemical as it was provocative, Osborne's book was about "two major conflicts." One was the Second World War, which demonstrated the violence and destruction humans could inflict on one another. The other was "man's conflict with nature," which, Osborne wrote, "contains potentialities of ultimate disaster greater even than would follow the misuse of atomic power."[39] However sharp these critiques were of postwar America, they did not generate enough popular support to engender a dramatic shift in postwar consumerism or a change in the technological apparatus that undergirded postwar modernity. Yet these arguments would evolve over the next two decades in the work of other critics of modern technologies, including a popular, but little-known, nature writer who would dramatically change the DDT story.

Despite Vogt's advocacy for biological processes, Mexican leaders encouraged the use of agricultural chemicals. The Mexican government offered economic assistance for agricultural development in the form of lower tariffs, state-financed loans, and government subsidies. In 1943, the Ávila Camacho administration removed trade tariffs to lower the costs of U.S.-made chemical pesticides.[40] Then, in 1945, Ávila Camacho used federal dollars to underwrite domestic chemical production companies, many of which were simple formulation plants that mixed American-made chemicals into finished products for market. He also established the state-controlled chemical production facility Fertimex, which produced chemical fertilizers and pesticides. Even though Mexican farmers might have rejected the Rockefeller Foundation's program or refused the entire technological package because they could not afford it, the cost, availability, and effectiveness of agricultural pesticides enabled farmers to adopt at least part of the program technology without much difficulty.

Despite these efforts, shortly after the war, DDT remained hard to come by overseas. While U.S. production of the chemical soared after the war, its distribution in Mexico was slow and haphazard. In 1947, the IHD director in Mexico, Wilbur Downs, claimed that DDT was "very scarce . . . and can be reckoned virtually as gold dust."[41] He decried the "exorbitant prices" of DDT, claiming that the "whole question of handling DDT and DDT products in Mexico [had] been completely unplanned and unregulated in the higher circles."[42]

As the head of the Rockefeller Foundation's malaria eradication efforts in Mexico, Downs was more preoccupied with the DDT situation in Mexico than most. Reporting to the foundation's New York offices, Downs proclaimed, "Having DDT on hand, over and above our estimated needs, gives me a very comfortable feeling."[43] In 1948, he tried to convince his Mexican counterparts to establish a central purchasing agency for DDT in order to insure low-cost availability of the product, but little was done to centralize the local supply of DDT.[44]

Nevertheless, because of the political importance of the Rockefeller Foundation's programs in Mexico, DDT was available from multiple sources. During the Second World War, Roosevelt, fearful of Mexico's anti-American sentiment and political support for Nazi Germany, created the Institute of Inter-American Affairs (IIAA) to improve relations between the two nations.[45] The agency, headed by Nelson Rockefeller (who was only tangentially connected to the foundation), urged U.S. business interests to advertise pro-American messages in Mexico and promoted a Mexican nationalism friendly to the United States. At the same time, the IIAA was interested in promoting health. In 1948, the IIAA offered to supply DDT and spray equipment, leading Downs to comment that the move would "be a real step forward."[46]

The problem of availability was compounded by research on DDT's viability as an agricultural pesticide. Initial results were troubling. In 1947, the former state chemist for California, Alvin J. Cox, wrote a two-part article on the use of DDT in agriculture for *Agricultural Chemicals*. "The research is vast," Cox wrote, and "in some cases the conclusions are contradictory." There was "considerable apprehension" that DDT "might kill off the parasitic and predaceous insects which normally would help to hold injurious species under control." Another report indicated, "DDT is not discriminating; along with the bad, it kills the good insects, including the lady beetles and honey bees."[47] To most agricultural experts, the "miracle" of DDT clearly came at a significant cost. While many were willing to overlook the ecological hazards of using the pesticide, the consequences of its use could not be neglected, particularly by agricultural scientists who regularly read and published in the leading journals, including the scientists affiliated with MAP.

Despite these well-known dangers, DDT's potential as a cheap and effective form of pest control made experts enthusiastic supporters of the pesticide for agricultural use. MAP researchers were no different. "Of the new insecticides tested," the Rockefeller Foundation noted in its year-end report in 1947, "none proved equal to DDT."[48] Another report suggested, "DDT is cheaper [and] it should be put into general use for the control of flies and other insects."[49] Writing in *Agricultural Chemicals*, Harrar observed, "Total quantities of pesticides are small, but these are increasing. A number of small mixing plants have been established . . . [and] the willingness of the producers of insecticides and fungicides to provide experimental quantities of their products has been a great stimulus to this development." He later concluded, "It is expected that the use of agricultural chemicals will be an accepted part of the production patterns throughout Mexico with resultant economic benefits."[50]

The foundation's research was set against a changing chemical landscape, as the agricultural changes in the United States had been.[51] After the war, the U.S. chemical industry introduced a series of new synthetic pesticides, including aldrin (1947), dieldrin (1948), toxaphen (1948), and endrin (1951). Like DDT, these synthetic chemicals, also known as organochlorides or chlorinated hydrocarbons, were broad-spectrum pesticides that "persisted" once applied. Although somewhat more hazardous than DDT, they were less acutely toxic than another group of chemicals that entered the marketplace.

Based on a derivative of a German nerve gas, organophosphates, namely parathion and malathion, proved to be effective killers of insect pests. They were also extremely toxic to humans and other animals.[52] Parathion, for example, is nearly twenty times more toxic than DDT, and a small drop absorbed through the skin can be fatal. Despite these dangers, one of the benefits of using organophosphates is that they break down rather quickly once applied, therefore reducing the long-term risks associated with the use of chlorinated hydrocarbons.

Another group of pesticides brought together the best and the worst of organochlorides and organophosphates. Derived from carbamic acid, carbamates disrupted the nervous systems of insects (as organophosphates did) but were less toxic to mammals (as organochlorides were). However, just as DDT was, they were broad-spectrum pesticides that killed beneficial insects like bees as well as insect "pests." First introduced in 1956 under the name "carbaryl," these chemicals became an extremely popular form of insect control in the United States, especially for home lawn and garden care.

MAP officials negotiated this new chemical terrain with great enthusiasm, but also much trepidation. These new agricultural chemicals introduced new risks, requiring the Rockefeller Foundation to consider an assortment of issues, including potential poisoning or hazards to other insect species. They were also in line with the foundation's belief that the Mexican project was just the gateway to a larger agricultural revolution. Writing to Elvin Stakman, Frank Hanson, the assistant director of the Rockefeller Foundation's Natural Science Division, commented, "I feel that any future program in agriculture in Latin America will depend almost wholly upon the degree of success and the results which come out of our Mexican work during its first decade."[53] This new chemical terrain required much consideration for the potential hazards to humans, and ultimately for the hazards to the agricultural development program itself.

Initially, DDT served as an important agricultural pesticide because of its low human toxicity and general effectiveness. However, as the chemical options expanded with the introduction of organophosphates, MAP officials continuously reassessed their recommendations for crop protection, balancing agricultural benefits against the risk to human bodies. One report indicated that parathion showed "remarkable efficacy" but was "an expensive and hazardous insecticide to use."[54] Chemicals "less dangerous than Parathion are being tested," the foundation suggested, "in order to secure a better insecticide."[55] Better, in this case, certainly meant less dangerous. Another report stated, "Parathion and dieldrin give excellent control at very low dosages," yet MAP did not recommend these pesticides for use "in Mexico" because of "their toxicity to human beings or their unavailability." It did recommend DDT, however, "even though somewhat higher concentrations [were] required." The same report later concluded, "Parathion is much more effective against the leafhoppers than either DDT or toxaphen but the hazards involved in the preparation and use of this chemical makes it unsafe for farmers to use. Therefore, it is recommended that DDT be applied."[56]

Undoubtedly, incidents of accidental poisoning created a problem for MAP and the agricultural model it promoted. From the Rockefeller Foundation's perspective, they could have catastrophic consequences, since they had the potential to endanger the program. As an institution working in a foreign land, the foundation depended on the health of Mexican agricultural workers for the success of the program. Ready access to toxic substances made the likelihood of accidental poisoning that much greater.

For example, in 1946, Rockefeller Foundation adviser Paul Mangelsdorf reported that an unnamed agricultural laborer working in an experimental field near Cuernavaca, Mexico (approximately fifty-two miles south of Mexico City), died unexpectedly. Local residents attributed his death to chemical poisoning, although the official cause of death remained unexplained. For Mangelsdorf, however, the situation had the making of "a major disaster," which could have "jeopardized the entire Rockefeller Program." The consequences of modern agriculture seemed to have been realized, a lifeless body amid the crops of the future. Mangelsdorf's report did not reflect on the human tragedy unfolding before him or on the dangers of agricultural labor. Rather, his observations focused on the political ramifications of this death. Not only were local farmers and laborers outraged about the death; other questions were raised: What would Mexican leaders think about the situation? What would the worker's death mean to the future prospects of agricultural development and the Rockefeller Foundation's development project in Mexico? The poisoning, Mangelsdorf believed, had the potential to compromise the "Foundation's future effectiveness in Mexico," if not the dreams of spreading the program globally.

To Mangelsdorf's relief, however, his traveling companion, a Mexican agricultural official, took "prompt and appropriate steps to safeguard the Foundation's prestige." Informing an angry crowd that the Mexican Department of Agriculture would take "the entire blame" for the incident, the agricultural official was able to "quell the disturbance." For Mangelsdorf, it was a fortunate turn of events. The foundation's work would continue unabated.[57]

However, that same year, *El Universal* reported on the death of a thirteen-month-old whose accidental ingestion of DDT produced "explosive currents" from the child's body. Similar cases began to appear throughout the nation, which greatly disturbed the Rockefeller Foundation. Even today, the scope of the problem is not entirely understood. Many cases of pesticide poisoning often go unreported, either because they are misdiagnosed or because the populations most at risk from poisoning are poor and do not have access to proper medical care.[58]

These tragic events revealed the complicated nature of the foundation's program and its impact on health. Yet MAP officials were committed to creating a modern agricultural system in Mexico, regardless of the cost and consequences. And because of its devotion to the promotion of agricultural technologies and its growing influence in the field of agricultural development, MAP faced a set of challenges that shaped its work in Mexico and throughout the world.

Pesticides posed one of the greatest challenges because of their high risk. Unlike other technological inputs that were safer, although not necessarily ecologically sound, like hybrid seeds, tractors, and chemical fertilizers, pesticides had the potential to undermine the basic logic of agricultural development. Elvin Stakman claimed, "We've got to make a choice." Not only did the foundation have "to be concerned about the direct effects" of pesticides, but it "had to use them with extreme care." These were, indeed, the dangers the foundation faced in promoting its version of agricultural progress. Stakman reiterated, perhaps in a more telling statement, the general disposition the foundation brought to Mexico and the "problem" it discovered in working with "foreign" peoples. He remarked:

I think that under certain circumstances some of these chemicals can be used safely. Now, in the old days we tried hard to be as sure as we could that there was no cumulative damage or hidden damage, because we were responsible and because we were recommending the use of these things to non-scientists, people who depend upon us. But for a very intelligent person, if you say to him for example, "Don't put this spray onto apples or plums or any fruit during the last two sprayings; use something else, because this might be deleterious,"—if you're dealing with a very intelligent person, you could depend upon him to do that. The trouble is, you're dealing with a lot of people who are not experienced in the sort of thing and don't use discretion.[59]

Stakman's claim that his Mexican counterparts were "not experienced" reflected the Rockefeller Foundation's belief in its own agricultural expertise—the belief that it had the knowledge to make wise decisions. And, perhaps more importantly, it reiterated the impact the chemical marketplace had on the agricultural landscape and the dangers of working in an increasingly toxic environment. Alvin Cox said it more directly when he wrote in 1947, "We are living in a chemical age, and it is to be expected that pests will be controlled more and more by synthetic organic insecticides."[60]

As the global market for agricultural chemicals expanded during the 1950s and 1960s, the growing market promised not only higher yields, but also increased health risks around the world. While many people grew healthier with an abundant food supply, others faced the adverse affects of increased pesticide use. Modern agriculture had its risks, but those risks were a price many countries, including Mexico and the United States, were willing to pay. Over time, these risks would be assumed by some of the poorest and most politically marginal people in the world.

To educate Mexican farmers about pesticides and modern agriculture, the Rockefeller Foundation and the Mexican Department of Agriculture developed a series of outreach programs. Agricultural films, demonstrations, and extension bulletins were important components of the foundation's extension efforts. By 1962, the foundation had produced sixteen films. Distributed throughout Mexico, "the films reached nearly 130,000 farmers" and encouraged viewers to adopt the foundation's technical expertise.[61] Only one film, however, focused on pesticides. *Proteja su cosecha* (*Protect Your Harvest*) showed viewers how to reap the economic rewards from the efficient and safe deployment of agricultural pesticides. MAP scientists also held a number of agricultural demonstrations throughout Mexico as a way to promote its program face-to-face. The Vera Cruz newspaper, *El Dictamen*, reported that these demonstrations offered "the most objective teaching" through which farmers and ranchers could learn more about "the experimental plots" and "appreciate the advancements and techniques obtained in the research centers" while taking advantage of the "opportunity to consult freely on special problems." The purpose of these events, of course, was to introduce farmers to "modern agricultural practice," including a "better system of battling insects."[62]

While these extension efforts were significant, they were limited in the number of people they reached. To expand its reach into the Mexican countryside, the Rockefeller Foundation distributed agricultural bulletins that promoted the technical achievements of its program. The type of information conveyed is reflected in some of the titles: *Proteja su cosecha* (*Protect Your Harvest*), *Abonar su maíz y coseche más* (*Fertilize Your Corn and Harvest More*), *Buscando una agricultura mejor* (*Searching for a Better Agriculture*), and *La ciencia agrícola impulsa al Noroeste* (*Agricultural Science Stimulates the Northwest*).[63] Many of these bulletins described how pesticides were sold—as prepared dusts, concentrated dusts, or liquid concentrations. The bulletins instructed farmers on how to prepare pesticides for use—to apply pesticides directly or disperse in water. And the bulletins described when and how to use pesticides on crops. Most also came with a warning. In one bulletin, farmers were told that "each year several people have been made ill and some have died due to exposure to insecticides or to ingestion of the same. These accidents should not occur if precautions have been made when preparing and applying the insecticides." Pesticides, the bulletins reminded farmers, should be used "with care."[64] The problem was that many farmers did not have the ability to read, thus making the bulletins basically worthless.[65]

Despite the Rockefeller Foundation's efforts to control the use and flow of pesticides and to protect farmers from undo harm, its experimental work engendered a fundamental shift in the marketing of synthetic chemicals in Mexico. Not only did the Mexican government encourage the consumption of modern pesticides, but U.S. chemical producers understood the importance of the foundation's efforts in Mexico and began to forge closer relations with MAP officials. The Mexican newspaper *Excelsior* reported on an experimental project with a new Shell Chemical Corporation herbicide, "D-D." Shell representative A. J. Garon indicated that the tests were the first experiments of D-D in Mexico and "owed their success . . . to the cooperation of the members of the Rockefeller Foundation."[66] In 1951, another Shell representative asked J. G. Harrar for information concerning the pesticides used in Mexico, since the company was considering a permanent move into the Mexican chemical market.[67] In 1953, Dr. W. H. Tinsdale, from the pesticide division of DuPont Chemical, inquired whether he could visit MAP to survey the work being conducted in Mexico.[68] One Dow Chemical executive commented, "The Mexican government controlled and isolated for itself basically all the petrochemical business . . . it's a hell of a market to build and get into."[69] Over the next two decades, however, U.S. multinationals increasingly moved in and dominated the Mexican market due to government intervention and investment in modern agriculture.[70]

The Mexican government's decisions were transformative. Whether growing crops for subsistence or for market, Mexican farmers had greater access to cheap pesticides during the 1940s and 1950s. One survey indicated that

Mexican "imports of insecticides increased from 500 metric tons in 1939 to 3,200 tons in 1949 and to 11,900 tons in 1950."[71] By 1955, over 33,000 tons of pesticides were exported annually to Mexico from European and American suppliers.[72] Domestic production increased as well. Between 1950 and 1955, Mexican manufacturers increased production from 2,500 tons to over 70,000 tons. Total pesticide consumption in Mexico swelled from 14,100 tons in 1950 to over 113,000 tons by 1960.[73]

In 1956, William Cobb, a staff writer for the Rockefeller Foundation, produced a brief history of MAP and the changes the program produced. Writing about the Mexican agricultural landscape prior to the foundation's arrival, Cobb described an agricultural system fraught with inefficiencies and incapable of dealing with insect pests with scientific proficiency. Cobb wrote, "In those innocent days science had not erected the defenses of chemical protection and entomological control which stand between the modern farmer and his natural enemies."[74] Cobb's use of the term "modern farmer" clearly characterizes the scientific and technological approach to agricultural development that the foundation sought to impose on the Mexican landscape. In the span of thirteen years, the foundation engineered a dramatic change within Mexico. For better or worse, the "innocent days" had been replaced through a scientific transformation that made nature an "enemy" to be kept in check with chemical pesticides.

The Rockefeller Foundation's work in Mexico also raised the possibilities of increasing agricultural yield globally. In the years after the war, concern over food shortages around the world grew louder. "Reserve stocks of food," the U.N.'s Food and Agricultural Organization proclaimed in 1948, were "at a low ebb." Serious crop failures would "bring widespread suffering" and "greater deprivation and unrest."[75] Fear of unrest led many U.S. policymakers to conclude that food production was necessary to ensure political stability abroad. MAP, therefore, was important on many levels. In 1953, Stakman spoke of the "world-wide importance of the program," claiming, among other things, that the foundation's work in Mexico was "looked on as [a] model project."[76]

These sentiments were expressed two years earlier in a report titled, "The World Food Problem." Written by Richard Bradfield, Elvin Stakman, Paul Mangelsdorf, and Warren Weaver, director of the Rockefeller Foundation's Natural Science Division, the report drew closer ties between agricultural development and security. As Lord Boyd-Orr had done in his advocacy for economic development in Sardinia and the political context within which these development models would emerge, the report connected the foundation's work to larger political questions of global food production. "The

time is now ripe, in places possibly over-ripe, for sharing some of our technical knowledge with these people. Appropriate action now may help them to
attain by evolution the improvements, including those in agriculture, which
otherwise may have to come by revolution."[77] The report provided a blueprint
for agricultural assistance on a global scale, calling attention to the importance of food production in an increasingly dangerous world. It is clear that
the foundation also played a greater role in shaping the political and ideological case for agricultural development based on the MAP model.[78]

If judged solely by increased yields, the agricultural model developed by the
foundation seemed to be an overwhelming success. By 1954, Mexico reported
the largest output of wheat in its history. And instead of importing grain to
make up for critical shortages in production, Mexico, for the first time in its
modern history, became a wheat exporter.[79] Although recent research suggests that the nutrition of basic grains may have been overlooked in favor
of increased yields,[80] for all those involved in the development project, crop
surpluses represented a dramatic improvement.

The chief architect of the Rockefeller Foundation's wheat program, Norman Borlaug, emerged as a dominant figure in the development project, and
he is often referred to as the "father of the Green Revolution." Working in
the irrigated plains of the Yaqui Valley in the northwestern state of Sonora,
Borlaug developed a hybrid variety of dwarf wheat that was capable of withstanding diseases such as wheat rust. And because the stems were shorter
and thicker, they were able to support bigger heads, thus generating larger
yields per acre. The downside of Borlaug's hybrid wheat was that it required
enormous amounts of chemical fertilizers, pesticides, and water to flourish.
This method of farming was capital intensive and technically sophisticated
agriculture. And for those engaged in producing crops, it was increasingly
dangerous. But because Borlaug's seeds generated enormous crop yields,
they would soon become the basis of the Green Revolution.

Borlaug claimed that in the span of ten years, a system of agriculture that
dated "back more than three thousand years" had been replaced by the "most
modern tractors and self-propelled combines of the twentieth century." He
wrote, "The only implements used in wheat cultivation were the Egyptian
plow, and the hand sickle. Today virtually all of the wheat plantings . . . have
been mechanized. Tractors, plows, disc harrows, land leveling equipment,
grain drill and self-propelled combines are employed on most of the wheat
farm operations."[81] In little more than a decade, the biblical landscape Henry
Wallace described had been transformed, and nearly 3,000 years of history
had been transcended by modern technologies and scientific know-how.

The celebrated success of the Rockefeller Foundation's program was not lost on those back home. An editorial from 1955 in a Beaumont, Texas, newspaper stated, "Surely, there is no better way for any American institution to create stronger ties of friendship and good will between people of the United States and the people of Latin America."[82] Similarly, the *Oregon Journal* stated, "Here is a real 'good neighbor' program — helping the other man live better through his own efforts and industry."[83] This was certainly the kind of response to American foreign investment that Henry Luce envisioned, even though *Life*, *Time*, and *Fortune* were remarkably silent about Mexico's agricultural bounty.

The Rockefeller Foundation's success in Mexico also served as the basis of President Truman's Point Four program. By offering technical assistance to developing nations, Truman believed that improvements to agricultural production and health would win the hearts and minds of civilian populations worldwide. Speaking at the Centennial Celebration of the American Association for the Advancement of Science, a gathering at which Elvin Stakman became the new president of the association, Truman proclaimed, "Scientific research daily becomes more important to our agriculture, our industry, and our health." He continued, "American scientists must, like all the rest of our citizens, devote a part of their strength and skill to keeping the Nation strong."[84] Truman fully understood the political and geopolitical context in which science could serve the state.

For its part, the foundation attempted to steer clear from any overt connections to Point Four or to U.S. policy more broadly, valuing its reputation as an independent scientific organization, despite evidence to the contrary.[85] In the early 1950s, representatives from Point Four met on a number of occasions with the foundation's agricultural team, suggesting that MAP "should be the pattern for Point IV work in agriculture."[86] The two programs seemed to be comparable. Yet Harrar was extremely critical of the Point Four agricultural program as a technical enterprise. He claimed that the program was "large and inclusive," presenting "problems which may be difficult to solve."[87]

To be sure, the scale of the Point Four project overwhelmed the Rockefeller Foundation's program. The Point Four project was a global initiative with projects conducted in thirty-six countries on three continents: South America, Asia, and Africa. Most of these projects involved some form of agricultural development work. What Harrar and other foundation officials realized, however, was that agricultural technologies alone could not remedy the inefficiencies of the past. What was needed, they believed, was a long-term commitment to developing an economic infrastructure capable of handling the complexities of modern agriculture, complexities including seed distribution, developing irrigation networks, creating or stimulating markets for agricultural products, establishing scientific research centers, and insuring the supply of chemical fertilizers and pesticides. Despite the *New York Times* headline declaring "Point Four to Urge Capitalist Farms," Harrar believed that Point Four was committed only to short-term goals that were largely political rather than humanitarian.[88]

Within Mexico, the story of agricultural development unfolded in multiple ways. To be sure, Mexican elites, scientists, and farmers were not passive actors in this agricultural transformation. During the Ávila Camacho administration, the Mexican political elite embarked on a set of policies that radically, and dangerously, transformed the Mexican countryside. Subsidizing the production and consumption of agricultural chemicals encouraged large and small farmers to adopt the tools of "modern agriculture" without guidance or concern. The Mexican government funded massive dam projects, including the Papaloapan project in the state of Vera Cruz, a project which many referred to as the "tropical TVA" because it supported large-scale agriculture.[89] In some cases, the nationalistic call to produce food for a modernizing nation overwhelmed the assumed risk to humans. In a 1953 telegram to Mexican president Adolfo Ruiz Cortines, the League of Agricultural Communities of Jalisco and the Federation of Small Property Holders offered their "profound thanks" for the "opportune credit" and "ample technical direction" that allowed them to "cultivate the land and to respond nobly and patriotically to [the president's] call for producing an abundance of basic foodstuffs for the nation."[90]

On the other hand, many rejected the Rockefeller Foundation's agricultural program, claiming, among other things, that the Mexican agricultural system was not as inefficient as the foundation believed.[91] Nor did Mexican farmers simply embrace the foundation's agricultural program blindly. Instead, many farmers and agricultural experts within Mexico observed the foundation's program with a critical eye. Some accepted the technological package wholeheartedly. Some rejected it outright. Others simply took elements of the entire package and adapted them to the existing agricultural system. For example, in his study on the Zapotec people in Oaxaca, Roberto González claimed, "It appears that fertilizer—the only Green Revolution artifact that has made a significant impact in the Sierra—has become indispensable."[92] It was clear, however, that despite these alternative approaches to Green Revolution technologies, the foundation's model of high-yield, mechanized agriculture became synonymous with the development of a modern nation-state.

So while the foundation embarked on its quest to modernize the unmodern, it did so with the best intention of helping people feed themselves. As the project unfolded, however, differences between expectations and reality became more pronounced, reflecting some of the underlying assumptions about modernization held by foundation officials. And despite the increased agricultural yields, large parts of Mexico were still mired in poverty. In 1955, President Ruiz Cortines said that Mexicans were still "living in shacks, with nothing to eat but tortillas, with no shoes, no education for their children, no hope. . . . Poverty, ignorance, and disease still plague many of our countrymen."[93] The agricultural bounty sowed by the Rockefeller Foundation could not deal with the structural issues of poverty or simply erase the past. And in many cases the foundation's projects may have contributed to the growing inequity between the wealthy and the poor.

On a global scale, the promise of high-yield agriculture and Green Revolution technologies generated enormous interest among developing and developed nations. During the early 1950s, the Rockefeller Foundation spread its agricultural development program to Colombia (1950), Chile (1952), and India (1955). Elvin Stakman recalled, "The Mexican program, in short, was the multiple mother of the agricultural revolutions that are under way in many countries of the world, especially of course, the underdeveloped countries, including some of the most populous."[94]

Through its long-term commitment to agricultural modernization, the Rockefeller Foundation created a vast network of institutional relationships, many of which engendered the further expansion of industrial agriculture in Mexico and around the world. These relationships helped bridge the gap dividing U.S. agricultural and chemical interests and Mexican agricultural officials and farmers. By paving new roads into the Mexican agricultural economy, MAP played an important role in facilitating the transfer of new technologies and chemicals across the border. These relationships would continue to develop in the ensuing years. And they would also become more political as the foundations of the American Century began to be debated more vigorously in the context of the Cold War.

As an agricultural pesticide, DDT showed tremendous promise within the development project. Because it was effective against insect pests and less toxic than other agricultural chemicals, DDT seemed like an almost perfect pesticide. However, the pesticide, as a historical actor, would begin to shape a history that challenged the linear narrative of development promoted by the foundation and other technological enthusiasts. The evolutionary realities of insect ecology also added to the development challenges as pesticide resistance led to a real problem. Indeed, the technological aspirations of development experts, shown in their unquestioned belief in DDT, would soon come face to face with ecology. And as development experts began a global assault against plagues and pestilence, the tension between technology and ecology would become increasingly obvious and difficult to ignore.

4

THE AGE OF WRECKERS AND EXTERMINATORS
Eradication in the Postwar World

Malaria eradication . . . stimulates development schemes.
—World Health Organization Report on the Second Regional
Conference on Malaria Eradication, May 20, 1959

In 1934, the noted urban critic Lewis Mumford released *Technics and Civilization*, the first of a series of classic studies on the connections between human societies and the technologies those societies developed. In *Technics and Civilization*, Mumford questioned the prevailing interpretations of technological change, suggesting that the social, cultural, political, and economic upheavals that materialized during the nineteenth century were not necessarily new, but were, in fact, part of a longer process rooted in medieval times. Mumford also suggested that the technological and scientific processes that emerged at the beginning of the twentieth century created something unique; this was a period he called the "neotechnic phase." Scientists, with an eye toward understanding all aspects of the natural and man-made world, Mumford argued, would change the twentieth century in ways unforeseen in previous centuries. "The concepts of science," he wrote, "hitherto associated largely with the cosmic, the inorganic, the 'mechanical' were now applied to every phase of human experience and every manifestation of life."[1] Indeed, science, as we have seen, was not confined to the laboratory, but was increasingly part of what Henri Lefebvre called "the everydayness of life." And as proponents of the American Century envisioned, technological science would become part of the everydayness of life within and, perhaps more importantly, beyond U.S. borders.

Mumford would go on to have an illustrious career as a writer and social critic. His books, *The Culture of Cities* (1938), *The Condition of Man* (1944), and *Cities in History* (1961), would challenge readers to grapple with the complexities of the modern world. Both a promoter of modern technologies and a critic of them, Mumford, for much of his life, straddled an intellectual fence that made him one of the unique voices in the twentieth century. While the early twentieth century was one of wild possibilities, the postwar years, for Mumford, were ones of disillusionment and unease. From atomic energy to the destructive capacity of the bulldozer, Mumford was aghast at the transformation of the American landscape. His last major work, the two-volume *The Myth of the Machine* (1971, 1974), reflected his disenchantment with technological modernity.

Amid this postwar transformation, in 1958, Mumford wrote a letter to Jane Jacobs, a young architectural writer who shared many of his concerns about the modern world. Although Mumford would later be critical of Jacobs, especially after the publication of her path-breaking book, *The Death and Life of Great American Cities* (1962), he found Jacobs's early writing on cities thought provoking. The letter reveals Mumford's fondness for Jacobs, while expressing his growing anxiety about the postwar world. "These are days, dear Jane Jacobs," Mumford wrote, "when we have reason to fear the worst, both from City Hall and from the White House. . . . I suppose when I suggest that if anything survives this age it will be known, retrospectively, as the age of the wreckers and exterminators."[2]

As many of Mumford's pronouncements did, his letter to Jacobs struck a particularly strident chord. While his comments reflected the radical transformation of urban space under the auspices of urban renewal, Mumford captured the tenor of his time with remarkable insight. Indeed, in an effort to recast the world, wreckers and exterminators wanted to create the new by erasing history. Wreckers and exterminators believed they were eliminating the inefficiencies of the past in creating a "modern" society, one predicated on the efficiencies of science and technology. And despite Mumford's conclusion that the age of wreckers and exterminators was something particularly American, it was in fact increasingly global, largely perpetuated by proponents of the American Century.

Perhaps no concept captured Mumford's assessment of the postwar world better than eradication. Eradication was, for all intents and purposes, the guiding principle that framed the interaction between humans and insects. If anything, the Sardinian campaign confirmed the viability of mosquito eradication in the minds of many public health and government officials. Yet, more than a strategy against insects, eradication was an ideology based on the fundamental belief that humans and the technologies they constructed could effectively control the natural world. It was a belief system that saw insects as an enormous threat to human health and development and adhered to the promise of science and technology as the arbiters of the modern world. In practice, as well as theory, eradication engendered passion and an unquestioned faith in human technology. Fred Soper declared, speaking to the Wisconsin Anti-Tuberculosis Association, "Eradication requires conviction, dedication, even fanaticism, which cannot be generated without faith in the objective."[3]

How eradication became a central concept in the postwar period has often been an overlooked question. Precisely because eradication symbolized the transformative possibilities of the postwar world, as people like Soper and Mumford understood, it has become something of an ahistorical phenomenon. Seen through the lens of either public health or agricultural history, the scope and breadth of eradication as an idea and an approach to insect control has been subsumed by disciplinary boundaries. By examining a longer history of eradication, one that envelops both agriculture and public health, we can better understand the evolution of an idea that transcended any one nation and became a global phenomenon during the American Century.

AT THE END OF THE ERADICATION campaign in Sardinia, one health official commented, "The concept that malaria could be totally eliminated from an area is an old one."[4] Indeed, the concept of eradication was not a product of the postwar world, but part of the longer process of the changing relationship between animals and humans. During the nineteenth century, for instance, Americans eradicated a number of wild species without much forethought. The state-sponsored massacre of the American bison transformed the Great Plains, converting prairie grasslands into an agricultural breadbasket, while erasing any vestiges of a Native American past. The wholesale slaughter of millions of song and migratory birds, including the passenger pigeon, whose extinction in 1914 symbolized the human exploitation of the natural world, was based on the belief that nature was both inexhaustible and ripe for human consumption. The successful eradication of insects, however, proved elusive.

Perhaps because of the complexity of insect ecology and the sheer number of insects, control, much less eradication, seemed like an impossible dream. For most of the nineteenth century, U.S. farmers and entomologists relied on biological rather than chemical methods as a way to control insect populations. Even with the advent of Paris green in 1814, which became the most widely used chemical in the United States by 1900, the transition from bio-logical to chemical control of insects was not immediate, nor was it widely accepted. Indeed, early agricultural chemicals like Paris green or Bordeaux mixture (copper sulfate and lime) were costly and dangerous to use. Some-times with the aid of chemical pesticides, but often through biological means, public health and agricultural officials simply attempted to control insects, not eradicate them.

The dawn of the twentieth century reconfigured the concept of insect eradication. The medical discoveries of Carlos Finlay, Walter Reed, and Ron-ald Ross encouraged public health officials to explore the possibility of elimi-nating mosquito vectors through environmental, biological, and chemical methods. William Gorgas, the noted American physician, developed various sanitary measures, including drainage and fumigation, to reduce mosquito populations in Cuba and Panama between 1900 and 1903. Joseph Le Prince, a British-born sanitary engineer employed by the Isthmian Canal Commission in Panama, significantly changed public health strategies for mosquito-borne diseases.

Le Prince established malaria and yellow fever control efforts in Panama. Attempting to eradicate yellow fever by reducing "the number of Aëdes calopus to such a minority" that the transmission of disease would stop, Le Prince broadened the terrain on which the struggle against mosquitoes was fought. After extensive observation, Le Prince was convinced that in order to reduce mosquito populations, fieldworkers had to kill mosquitoes within buildings or dwellings, as well as in breeding areas. "The nearer the reduc-tion reaches complete eradication," Le Prince remarked in his influential book, *Mosquito Control in Panama* (1916), "the better especially in regard to the reintroduction of the disease at some future time."[5] Like Gorgas, Le Prince understood the intertwined ecologies of humans and mosquitoes, and this approach would "remain a bible for mosquito control workers for many years."[6] Le Prince also recognized that the massive environmental transfor-mation of Panama added to the mosquito problem there. The disruption of tropical ecosystems created new mosquito breeding grounds, while the influx of human labor made disease transmission more likely.[7] To this day, the American Society of Tropical Medicine and Hygiene awards the Joseph Augustin Le Prince Medal every three years for outstanding work in the field of malariology.

Similarly, public health officials in Florida launched an aggressive cam-paign to reduce malaria rates in an effort to encourage economic develop-ment throughout the state. In 1918, according to historian Gordon Patterson, the Florida Bureau of Engineering announced plans for the "eradication of mosquito breeding places" relying on drainage and oiling. Beginning in 1922, Miami and Tampa expanded "mosquito eradication" efforts, using a combination of techniques, including the use of chemical pesticides like Paris green. These efforts demonstrated the potential for mosquito control, which according to two Florida-based entomologists, was "more than a dream or idle fancy."[8]

As a unifying approach to malaria control, however, eradication was not a concept widely accepted. Ecological variables, not to mention the tremendous cost and manpower required for early eradication projects, challenged even the most ardent antimosquito advocates. Those less willing to embark on such projects saw the control of mosquitoes and disease transmission as the most useful means of combating insect-borne diseases. Yet what became increasingly clear during the first two decades of the twentieth century is that more and more sanitary engineers, public health officials, and local communities embraced the concept of eradication, even if its meaning had not yet been fully understood.

Nowhere is this more clear than in Walter Ernest Hardenburg's book *Mosquito Eradication* (1922). Hardenburg, an American engineer, had extensively traveled throughout much of South and Central America at the beginning of the century, earning him the moniker the "tropical tramp" from one of his contemporaries. While only in his early twenties, he wrote a number of notable pieces about his travels, including his influential book *The Putumayo— The Devil's Paradise—Travels in the Peruvian Amazon Region and an Account of the Atrocities Committed upon the Indians Therein* (1912).[9] A detailed account of the Putumayo region and the horrific treatment of native populations by the British rubber interest, the Peruvian Amazon Company, Hardenburg's observations were instrumental in a parliamentary inquiry into the company's practices, including slavery, rape, and torture. Hardenburg's experiences shaped his understanding of nature, labor, health, and capitalism, ideas that would influence his writing.

Indeed, Hardenburg's travels throughout the Amazon hardened his view of nature. Despite his disdain for the ways in which the Peruvian Amazon Company extracted rubber, Hardenburg believed the region was an untapped resource in need of a civilizing touch. "The whole Amazon Valley, when it should be opened up," he wrote, "will prove to be one of the most valuable parts of the earth's surface."[10]

Hardenburg's interest in development was also revealed in his book *Mosquito Eradication*, which was a comprehensive summary of existing research on mosquitoes and their habitats. Although a prescription for mosquito control more than eradication, Hardenburg's study, nevertheless, expounded on the struggle to tame nature for economic development. As many postwar eradication advocates did, Hardenburg understood malaria and diseased environments as both a public health and an economic issue. Writing about New Jersey, Hardenburg claimed, "The mere fact that mosquitoes are abundant in a region holds back development of that region."[11] Similarly, the economic dormancy of the American South was in large part, according to Hardenburg, a result of malaria and mosquitoes. Beyond U.S. borders, mosquitoes remained the "chief obstacles to man's conquest of the tropics." Through eradication, these problems could be "brushed aside" so "that great engineering works like railroads, canals, and other modern necessities of trade could be carried through to completion anywhere in the tropics."[12]

Hardenburg's vision of eradication was predicated on the same developmentalist model that would emerge during the postwar period. The processes of eradication, however, differed substantially. Hardenburg saw mosquitoes as a local problem. Although the "tropics" suggested a geographically broad region bound more by its proximity to the equator than by nations, he spent little time exploring this region in his book. Instead, Hardenburg focused on "towns" as the primary actors to institute, manage, and promote eradication projects. Often created in cooperation with the U.S. Public Health Service — a national institution established in 1902 to prevent the spread of contagious disease — the eradication projects Hardenburg began reflected his belief that "the average town can free itself from malaria and kindred mosquito-borne diseases and the annoyance of mosquitoes for much less than it costs to endure them."[13]

Although Hardenburg did not ignore the transborder migrations of mosquitoes, his advocacy of eradication at the local level recognized the political constraints that shaped public health projects. "Along with the marked increase in public appreciation of preventive medicine, that has developed in recent years in the United States, has come a recognition of the great importance of mosquito eradication, particularly in the South," Hardenburg claimed. "This has manifested itself in the launching of anti-mosquito campaigns in cities, towns, and villages in every Southern State, and it is a foregone conclusion that the success of these will spur other communities on to action."[14]

Moreover, Hardenburg believed eradication could be achieved through environmental and biological methods. His training and experiences as a sanitary engineer certainly shaped his ideas, suggesting that the manipulation of nature could have profound impacts within the public health sector.

For Hardenburg, drainage represented "one of the most reliable and permanent methods of mosquito control." As for biological control, Hardenburg argued, "There is no doubt that fish control, where it is applicable, is one of the cheapest and most satisfactory methods of fighting the mosquito." And if all else failed, he claimed, the "oiling of water surfaces should be considered as merely supplementary to drainage and other measures of mosquito control."[15] The oiling of water surfaces created a barrier between air and water, preventing larvae from developing into adult mosquitoes. Last on his list were chemical larvicides. Hardenburg observed that larvicides "may be utilized to advantage under some circumstances." "However," he stipulated, "it should be remembered that they are often poisonous, less effective and more costly than oil." Paris green only received Hardenburg's passing interest, primarily because of its potential risks.

Because Hardenburg's methods were rooted in local environments, and shaped by local conditions and local hazards, the way local communities went about eradication differed substantially, depending on specific environmental, economic, and political circumstances. In this regard, mosquito eradication was mainly about changing the environmental conditions to prevent mosquito-breeding areas rather than deploying chemical pesticides to kill adult mosquitoes.

Ten years later, Fred Soper's model of eradication changed this calculus. Because his method was based on the extensive use of Paris green, Soper fundamentally reframed the history of the interaction of humans and mosquitoes. Although not alone in his advocacy and zeal for eradication, Soper was, according to Malcolm Gladwell, "the General Patton of entomology."[16] His work in Brazil during the 1930s and 1940s and in the Mediterranean during the Second World War established a new model for public health. And with the advent of DDT, the application of chemicals on diverse geographies transformed the delivery of public health services for millions of people. Indeed, the "success" of the Sardinian campaign demonstrated the power of DDT as the primary weapon against mosquitoes, leading Soper to proclaim in 1951 that "today [DDT] is recognized as the key to the control of insect-borne diseases."[17] Moreover, the Sardinian campaign was significant not simply because it was the first large-scale eradication campaign that relied on DDT. The campaign was important mostly because it defied nature. Indeed, by eradicating the *Anopheles labranchiae*—Sardinia's native mosquito species—public health experts believed they had removed one of the last barriers to acceptance of the eradication logic. The challenges of controlling or exterminating a domestic or, as Raymond Shannon suggested when he visited in the Orlando laboratory in 1943, a "superdomestic" anopheles vector, were much greater, since the species was interconnected with a specific ecosystem in ways that an invasive species was not. Although earlier eradication models sought the removal of native mosquito species, especially along the American Gulf Coast, these eradication efforts tended to be more localized. DDT enabled public health experts, government officials, and international aid organizations to envision a world without mosquitoes, regardless of place and history.

Although years in the making, the transition to DDT as a single weapon of choice not only revolutionized public health, but marginalized other environmental or biological methods of mosquito control.[18] And rather than being a local concern, the focus on disease and eradication echoed far and wide. In 1950, Paul Russell, the esteemed malariologist, wrote, "Nothing on earth is more international than disease. Perhaps someday public health will also be universal."[19] Having just completed a three-year stint as the chair of the World Health Organization's Expert Committee on Malaria, Russell understood the global implications of disease, believing disease sabotaged the economic potential of the developing world.

Unlike Soper, however, Russell held that DDT should be used within a much more broad-based strategy, similar to the anti-mosquito approach developed by Reed, Gorgas, and Le Prince, arguing that environmental, therapeutic, and cultural practices should be an integral part of any eradication campaign. Others within the Rockefeller Foundation shared these beliefs. Wilbur Downs, for example, argued for a more broad-based approach, emphasizing control over eradication, multilateral techniques over unilateral techniques, and local initiatives aided by a well-planned national campaign. Downs claimed that a more sophisticated approach "can demonstrate that there are several approaches to malaria control. This can serve as a brake to the over-enthusiastic acceptance of DDT house-spraying as the beginning and end of malaria control work."[20] Yet Soper's advocacy, as well as DDT's low cost, relative ease of use, and effectiveness, led many to favor chemicals over other means of controlling mosquitoes. The chemical age had, indeed, come to fruition. And with it, the transfer and movement of DDT and other chemical agents would become a critical part of a global crusade against insect-borne diseases and other "plagues."

Despite his hesitation about DDT, Russell embraced the chemical's global possibilities. "For the first time," he wrote in his 1955 book, Man's Mastery of Malaria, "it is economically feasible for nations, however underdeveloped and whatever the climate, to banish malaria completely from their borders."[21]

For Russell the potential for eradication came not simply from the end of a spray gun, but from institutional structures established to support eradication on a global scale. After years of discussion and debate, the World Health Organization launched its Global Malaria Eradication Programme in 1955, much to the delight of many public health officials. Infecting nearly 750 million people worldwide, and leading to an annual death rate of 7.5 million per year, by 1950 malaria had killed more people than any other disease.[22] "The eradication of malaria from the world as a public health problem," the organization announced, "is a basic objective of WHO. Thanks to the remarkable properties of DDT, a number of national health departments, often assisted by WHO, UNICEF, or USA bilateral aid programmes, have made truly amazing progress towards this objective. Until recently the goal had seemed to become increasingly attainable."[23] For many in the public health community, the time was clearly ripe to eradicate mosquitoes and malaria from the world. Almost immediately, the Eisenhower administration offered $5 million, "without which," historian Amy Staples has suggested, "the campaign might have foundered for lack of funds."[24] In the years to follow, the United States increased its support dramatically, appropriating nearly $512 million over a five-year period beginning in 1958.

The global eradication campaign was based on four steps: preparation, attack, consolidation, and maintenance. Participating nations would design and carry out eradication efforts themselves while consulting with WHO experts. The preparation stage was to ensure that participating nations understood the scope of the malaria problem within their borders and that each nation had created the administrative framework to oversee the eradication program. Designed to last three to five years, the attack stage deployed health workers far and wide to spray homes, businesses, and public buildings with DDT. The consolidation phase required careful surveillance of targeted areas to ensure suppression of the disease and mosquito vector. If the desired results were not achieved, additional spraying would be compulsory. Finally, only when malaria eradication had been achieved would a participating nation reach the maintenance stage. Many believed that this prescription, regardless of history, ecology, and national borders, would erase the tremendous burdens and costs of malaria around the world. And it was all designed to occur within the span of a decade.

For Russell, Soper, and other like-minded malariologists, the appeal and possibilities of eradication were nearly universal. Variations in climate and geography, as well as the social, economic, and political disparities among nations were thought to be irrelevant in the contest against insect pests. Yet the desire for eradication was met with many foreseen and unforeseen obstacles, which created their own set of challenges as spray teams went into the field.[25] Moreover, the environmental impact from such a massive spray program was thought to be inconsequential, when in fact it left a considerable environmental "footprint."[26] Nature not only responded to malaria eradication, it also shaped its history.

Amid all the enthusiasm for DDT, the WHO's work for global malaria eradication continued despite mounting evidence of insect resistance. The experience in Sardinia certainly confirmed this. "The very fact that these rumours [of insect resistance to DDT] had arisen in regions where DDT had been utilized a year or a couple of years before," the WHO's Expert Committee on Malaria reported in 1947, "is rather disquieting, especially if we consider it in the light of some recent observations."[27] A year later, researchers indicated, "DDT-resistant populations of house-flies have developed. These findings should warn us to be on the alert in case any resistance of this kind should develop in the anopheline species, in order to be ready to replace DDT by other insecticides, if need be." In 1950, the WHO explained, "It is now apparent that in most areas there are strains of flies which are resistant not only to DDT, but also to other insecticides of the chlorinated hydrocarbon group."[28] By 1951, public health officials working in Greece found that mosquitoes had developed resistance.[29] And two members of the WHO Expert Advisory Panel on Insecticides commented in 1955, "The problem of the resistance of insects to insecticides is one of the most serious which has confronted the health services in recent years."[30]

For entomologists and public health officials, pesticide resistance was not an unforeseen phenomenon. In 1914, the San Jose scale, one of the major pests of citrus crops, demonstrated resistance to a lime-sulfur spray. Other signs of resistance became evident in the decades that followed, even though entomologists tended to turn a blind eye to the problem prior to the Second World War. Insect resistance became a global problem in the years after 1945, during postwar eradication and development efforts. Three years after launching the global eradication program, the WHO's Expert Committee on Malaria reported, "Should there be proof of resistance to both groups of chlorinated hydrocarbon insecticides, the Committee feels that, in the present state of knowledge, no residual insecticide now in current use, nor mixture of such insecticides can be recommended and mass chemotherapy should be put into operation without delay."[31] Each passing year, moreover, there were increased concerns about resistance. "A disturbing feature of the new records," the WHO reported in 1959, "is the apparent development of dual resistance (i.e., to both groups of the chlorinated hydrocarbons) in certain vectors of the Eastern Mediterranean Region."[32]

Elsewhere and over time, the problem of resistance continued to mount. Reports out of Latin America in the early 1960s suggested, "Problem areas in the Americas have been found usually to be due to a combination of factors. While double resistance (to both DDT and dieldrin) is the most important single factor, DDT usually continues to exert considerable suppression of transmission, although not sufficient to halt it entirely, even in the presence of resistant vectors."[33] The promise of eradication seemed to be crumbling just as quickly as it developed.

Insect resistance was a complex problem. But the evolutionary principles that resulted in resistance were rather straightforward. When DDT, or any other pesticide for that matter, was initially applied to an insect vector, the outcome reduced most of, but not the entire, insect population. This was due to either some genetic mutation that protected the insect from the chemical pesticide or some environmental factor that prevented a particular insect from coming into contact with lethal concentrations of the chemical. Whatever the reason, because the offspring of the living insects carried the genetic mutation or because, during the developmental stage, the insect came into contact with trace amounts of the pesticides, adult insects were more predisposed to resist the pesticide's toxicity. The progeny of these insects continued to evolve and adapt to changing environmental factors, thus producing "super insects." Additionally, these insect populations were more likely to grow at a faster rate, since many broad-spectrum pesticides attacked both the pest and its predator, a consideration that was particularly true with DDT. And because a population of prey may well outnumber its predators, "pests often become resistant before their natural enemies," resulting in a greater "pest problem" over time.[34]

Prior to 1946, only 12 cases of resistance were reported in the United States. Today more than 500 insects have become resistant to pesticides. And according to entomologist Marjorie Hoy, "Resistance to pesticides . . . is an evolutionary response to stress that will remain a persistent problem."[35]

Confronted with this evolutionary reality, one WHO official asked, "Can resistance be overcome?" And despite the cautionary tale houseflies and mosquitoes (as well as many other insects) were telling, many eradicators simply rearmed, rather than rethink their approach to the insect problem. "The only logical aim now, in the face of this problem," one WHO report advocated, "is complete eradication of malaria."[36] Rather than dissuade WHO officials and other eradication advocates to slow the deployment of DDT, pesticide resistance simply encouraged it. "If a country-wide malaria-control programme by residual spraying is systematically pushed to completion in a matter of seven or eight years," Emilio Pampana wrote in 1955, "there is not much chance of failure because of insect resistance to the toxicants." Pampana, an Italian-born physician, directed the WHO's malaria activities until 1958 and would later write "the textbook" of malaria eradication. He believed that the best remedy for resistance was the immediate and comprehensive use of DDT. For Pampana and others, the moment was at hand and time was of the essence. The global assault on the insect vector was "imperative . . . before all the species of malaria-carrying mosquitoes would develop resistance to the insecticides used."[37]

Some entomologists, however, argued otherwise, suggesting that eradication and DDT were not the magical elixirs their proponents suggested. In his 1955 study on the causes of resistance, Robert Metcalf concluded that resistance could be prevented "by the use of mixtures of toxicants with independent uncorrelated action or perhaps even better, the employment of alternated treatments with materials of independent uncorrelated action, . . . allowing the natural parasites and predators to eliminate the more resistant insects."[38] Another report indicated, "The epidemiology of malaria exhibits many contradictory aspects in different parts of the world," since "no species of anopheline vector has a world wide distribution" and because "different anopheline species vary widely in their bionomics."[39] "Residual spraying," another finding emphasized, "has not been universally successful in rapidly reducing malaria rates."[40] And a report from one field trial in the Belgian Congo on the aerial application of DDT indicated that "massive intervention sometimes risks upsetting a biological equilibrium favourable to the aims pursued."[41] Nevertheless, eradication as an ideology (and a practice) superseded individual concerns over the ecological complexity of the insect-disease axis.

Similarly, U.S. health officials and agriculturalists embraced the concept of eradication. Speaking before the 58th Annual Meeting of the American Association of Economic Entomologists in 1946, association president Clay Lyle told the assembled crowd, "Is not this an auspicious time for entomologists to launch determined campaigns for the complete extermination of some of the pests which have plagued man through the ages?" An enthusiastic supporter of chemical technologies, Lyle identified a number of insect species ripe for eradication: the gypsy moth (the eradication of which was already under way in Massachusetts), the housefly, and the horn fly. Lyle told his audience that DDT would ensure that there were "no flies in Idaho," and no cattle grubs, cattle lice, screwworms, or Argentine ants either. Capturing the tenor of the postwar euphoria about DDT and other chemical pesticides, Lyle concluded his remarks by quoting the famed nineteenth-century architect Daniel Burnham, telling his fellow entomologists to "make no little plans. They have no magic to stir men's blood."[42]

However, postwar eradication did not automatically presuppose the use of chemicals. Edward Knipling, the man who oversaw the Orlando laboratory during the Second World War, directed nationwide research on insects affecting livestock, man, households, and stored products for the Department of Agriculture. By 1953, Knipling was named director of the Entomology Research Division at the Agricultural Research Service (ARS) headquarters in Beltsville, Maryland. Under his leadership, the ARS launched an aggressive campaign to eradicate the screwworm from the American South. A parasitic insect that feeds on the live tissue of cattle, the screwworm caused enormous problems for ranchers, who saw the insect eat into their profits. Rather than advocate the eradication of the screwworm with chemicals, Knipling spearheaded a campaign that relied on the sterilization of the male screwworm using gamma radiation, a process that would interrupt the reproductive capacity of the insect. Operating out of the USDA's research station in Kerrville, Texas, Knipling's sterilization campaign produced favorable results. Another research project in Curaçao, the small Dutch island in the Antilles, also brought auspicious findings. Shortly thereafter, irradiated male screwworms were released throughout Texas and Florida. Initial results led to the expansion of the program, leading to the complete eradication of the screwworm within the United States by 1972.[43]

Yet it was another campaign in the American South that demonstrated the zeal and, perhaps more importantly, the futility of eradication: the fire ant wars from 1958 to 1982. Alarmed by the spread of the red imported fire ant (Solenopsis invicta), a particularly aggressive ant accidentally introduced into the United States in 1930, entomologists from the ARS began steps to eradicate the species. Despite the limited ecological damage caused by the fire ant, the ARS, with the help of local media outlets, national reports, and general anxiety over the furiousness of the ant species, targeted the ant as a pest that needed to be eradicated. More than any other insect, the fire ant came into the crosshairs of eradicators not because of some public health menace or dire threat to agricultural production, although the fire ant built mounds nearly four feet tall, but because they were dangerous "invaders." As historian Joshua Buhs argued in his study of the fire ant, "Pests are not born; they are made."[44]

Responding to this threat, the ARS deployed heptachlor, another chlorinated hydrocarbon. Reports indicated, however, that wildlife of "widely varied species" were found dead after application. While USDA researchers attributed these deaths to chemical pesticides, they also claimed, "While initial losses were relatively high, there are no indications that the use of insecticides will have a long-range effect on the population of wildlife in the area."[45] Beginning in 1964, however, mirex, another chlorinated hydrocarbon, was the chemical of choice. Unlike the heptachlor application, which was sprayed on about 10 percent of the fire ant range, largely within the area from Texas to Georgia, mirex was sprayed or used with a bait mixture on approximately half of the fire ant range by 1967.[46] Amid this intensifying chemical assault, not only was there evidence of "collateral damage"—large fish kills, reduction of beneficial insects, and ecological damage—but the range of the fire ant continued to expand.

Critics of the ARS's eradication efforts believed, as many of those who rejected the logic of urban renewal did, that experts misused their authority to support projects of dubious design. Rachel Carson was one such person. Her exposé on the excessive misuse of chemical pesticides, *Silent Spring* (1962), sounded the alarm about the ARS's narrow-minded effort to eradicate the fire ant. Over time, however, the chorus grew louder. According to one Texas resident, "It is obvious to me that the USDA is using this program for a very selfish and self serving purpose."[47] Eradication undermined a more complex understanding of the relationship between insects and humans. And knowledge developed at the local level and outside the boundaries of professional science was largely ignored. As they had in many respects in American society during the postwar years, experts often wielded tremendous power that regularly went unquestioned. Those in control of the spray gun dominated the public discourse about human health, environmental protection, and the role of chemicals in the world.

That does not suppose, however, that many scientists within the institutional ranks of the WHO or the ARS failed to recognize the ecology of disease and plant science. In fact, some of the most forceful advocates for the biological and cultural control of insects (rather than eradication) came from these institutions. Edward Knipling, for one, was not predisposed to reach for the spray gun as the first line of defense against insect pests. Instead, as demonstrated by his methods in the screwworm eradication campaign, he was committed to a much more broad approach to insect control. However, Knipling was unyielding in his defense of the ARS and the field of entomology. He was quite defiant toward unwanted or uninformed criticism and often dubious of nonspecialists.[48]

Under Knipling's leadership, the ARS was perhaps the most powerful agency within the United States supporting the chemical control and/or eradication of insect pests. At the height of the eradication campaigns, nearly 80 percent of the ARS budget was allocated for research on chemical controls. So even if Kipling's own research supported biological control of insect pests, the ARS devoted most of its financial resources to chemical controls. "Like demented physicians," historian Pete Daniel wrote, "ARS administrators prescribed chemicals indiscriminately, usually without full knowledge of the target pest and seldom with any thought of side effects."[49] The fire ant eradication project certainly confirmed the lack of foresight and ecological knowledge on the part of ARS officials. So while individually many scientists understood the complexity of the insect problem, within the institutional parameters of the ARS or the WHO, much of this nuance was lost with the zeal of eradication and "big plans." This tension characterized postwar eradication throughout the world. Individual reports often expressed caution or doubt about eradication. But set within a larger ideological framework, these individual concerns went largely unheeded.

As the concept of eradication captured the imaginations of people and institutions far and wide, the ethics of eradication and its impacts also began to resonate within and beyond the scientific community. Just as postwar eradication required "big plans," it also raised a number of "big questions" that reaffirmed and questioned the rationality of chemical technologies. On the one hand, malaria eradication had a profoundly positive impact on millions of people around the world, resulting in lower death rates throughout the 1950s and 1960s. Yet on the other hand, the resultant population increase created new problems.

The British physiologist and Nobel laureate Archibald Vivian (A. V.) Hill addressed the ethical dimensions of postwar development projects in his presidential speech before the British Society for the Advancement of Science in 1952. "In many parts of the world," he proclaimed, "improved sanitation, the avoidance of epidemics, the fighting of insect-borne disease, the lowering of infantile death-rates and the prolongation of the space of life have led to a vast increase of population." He added, "Nobody would dare say that steps to combat these diseases . . . should not be taken on the highest priority: but the consequence must be faced that a further [population] increase of a million people a year would result."

Concern over population growth was a logical outgrowth of malaria eradication. And it was this concern that led Hill to argue, "Thus, science, biological, medical, chemical and engineering, applied for motive of decent humanity entirely beyond reproach, with no objectionable secrecy, has led to a problem of the utmost public gravity which will require all resources of science, humanity and statesmanship for its solution." And perhaps responding to the Rockefeller Foundation's agricultural development model in Mexico, Hill told his audience, "It is *not* a question only of food; if a higher standard of life is to become universal, with education, communications, housing, reasonable amenities and public health, a far greater demand will be made on all such natural resources as power, chemicals, minerals, metals, water and wood." And because the basic tenet of the American Century was to spread the gospel of consumption, Hill submitted, "One is left wondering how long these [resources] can possibly take the strain. Could world supplies conceivably hold out if the present requirement, per head, in the United States, were multiplied in proportion to meet the same demand everywhere—even without any increase of present population; and if so, for how long?"[50]

Indeed, the American Century created a series of paradoxical questions that challenged the logic of eradication. Population growth, resource development, ecological change, consumption, and capitalism, all promoted under the guise of humanitarianism and expanding American interests abroad, underscored the significance and limits of U.S. science and technology. While the humanitarian ambitions of eradication were clearly delineated, few advocates of the American Century considered the long-term impacts of the global eradication program on the environment and human societies. This would change over time as fears of a global population explosion alarmed many policymakers concerned about the limits of the earth's resources and the geopolitical implications of hunger on a global scale. Meanwhile, malaria eradication continued unabated.

To reach the goal of eradication, the WHO required the full commitment of its member nations. "Unless the government gives to malaria eradication its full co-operation one hundred per cent," Emilio Pampana proclaimed, "efficiency cannot be reached, and this may mean that the spraying operations will have to be prolonged for another year."[51] Not only were concerns about DDT resistance mounting, but WHO administrators also understood the limits of nation-states to fulfill the WHO's global vision. No one nation could effectively eradicate malaria without neighboring countries' doing the same, since mosquitoes and disease often migrate across borders. Because of the migration of disease, the WHO persuaded countries to organize and coordinate eradication programs along border areas. And in some cases, the WHO developed regional programs, primarily along the western and southeastern African coasts, that brought together a number of countries under a single administrative director.[52] By building a cross-border network of scientists and scientific institutions, the WHO sought to forge alliances that transcended political borders in an increasingly divided world. By 1965, more than 100 countries participated in the WHO program.

From a U.S. perspective, however, national boundaries mattered for a much different reason. Not surprisingly, geopolitics and technological change were never far removed from American visions of eradication. In a 1959 letter describing the context and political impact of eradication, Paul Russell wrote, "In my opinion the success or failure of the world-wide Malaria Eradication program will depend in large measure on whether or not the U.S.A. is willing to continue and to increase its support. We have gained considerable good will in this program so far and definitely are ahead of the USSR in this field. I hope that we can stay ahead and reap the large reward in prestige that will come in a few years when a majority of the 76 countries now involved in the world-wide program successfully complete their project."[53]

That same year, Fred Soper spoke before the American Society of Tropical Medicine and Hygiene, claiming that communist countries had the upper hand in eradication because of their "interest in the productivity of their population and . . . relatively tight controls over their peoples." Soper believed that communist countries "may be expected to eradicate malaria as fast as or even faster than the free world, especially since most of these countries are not tropical." Although staunchly anticommunist, Soper may have had a soft spot for Soviet-style eradication because of the "discipline" it reflected. Indeed, of a Soviet project that began in 1952, observers noted, "three years later it became obvious that total malaria eradication from the country was possible by 1960."[54]

Despite its name, the Global Malaria Eradication Programme was far from global. Although much of Latin America, Europe, the Middle East, and Southeast Asia was awash with DDT, Africa was largely overlooked. While there were small experimental projects on the continent, WHO administrative support and U.S. funds never flowed freely into Africa. Strategically it was not as important as other hotspots of the Cold War, but many also believed that environmental concerns, as well as the political and structural upheavals of decolonization, would prevent the desired result. Faith in eradication required DDT but also a strong and committed national government capable of administering a four-stage program.

"To those who know Africa," one report suggested, "the idea of eradication of malaria there may seem Utopian."[55] Another report, from the program's general director, claimed, "The important point to be emphasized is that, excepting only tropical Africa, there are not in each continent malaria-cleared areas that demonstrate beyond doubt the economic and technical feasibility of malaria eradication by residual spraying."[56] The WHO commented in 1959 about its experimental work in Africa that although "several malaria eradication pilot projects have been tried, the results so far have been disappointing from the standpoint of total interruption of malaria transmission, even after 3–4 years of spraying." Aside from the "administrative shortcomings [and] lack of legislation and health education" that plagued other projects, "the principal barriers to malaria eradication in Africa," the WHO observed, "seem to be mainly attributed to lack of funds and lack of trained personnel."[57] The same could not be said for India, a country of strategic importance to the United States.

Beginning in 1953, the Indian government, with the assistance of the U.S. Point Four representatives, embarked on a malaria control program. By 1956, the program was converted "to an eradication programme." Five years later, the WHO summarized the campaign in a report that highlighted the scale and scope of the program, defined the difficulty of eradication, and recognized its strategic importance: "The initiation of a programme of such stupendous magnitude as the complete eradication of the world's greatest health problem by a country of over 400 million inhabitants with the economic difficulties found in India reveals a courage that is admirable." While Indian malariologists indicated that they were "cognizant that the programme demands a dedication unsurpassed in previous health undertakings," they were also aware that "problems not envisaged at the outset have arisen [and] that more such problems will arise." Resistance was one of the problems that "could greatly increase the technical and financial difficulties of the task." Another was the lack of cooperation or "refusal" of residual spraying on the part of many Indian citizens. Despite these problems, India was "committed to the eradication of malaria," and as was the case with most eradication projects, "there [was] no turning back." Fulfilling the goal of eradication "require[s] a unified command and a near-military precision in the operations," something that would be extremely problematic "in a federation of states."[58]

U.S. support for India's eradication campaign, moreover, fit within broader designs for the American Century. Although relations between India and the United States were often contentious, especially in light of Indian prime minister Nehru's intention to take a "third way" as a means to avoid or rewrite the bifurcation of the Cold War, the two countries had much in common, especially the belief in science and technology. According to historian George Perkovich, "To effect this third way, Nehru would rely on science and technology, including the scientific spirit, as valuable tools."[59] And in the quest to eradicate disease, both India and the United States recognized the value of DDT and the transformative power of science and technology. These connections became stronger as the agricultural revolution in Mexico emerged as a critical component of U.S. development efforts during the 1960s.

NATIONS, INTERNATIONAL AID INSTITUTIONS, and scientists were not the only ones shaping the history of DDT. As we have seen, the evolutionary processes that made insects resistant to certain chemicals affected how humans deployed pesticides. Moreover, the migration of DDT and other persistent hydrocarbons added to the increasingly intricate web connecting the human and nonhuman worlds. Indeed, like the evolutionary biology of insects, DDT had its own ecological story to tell, a story few cared to listen to.

As a chemical that persisted once it was released into the environment, DDT had long-term impacts most eradicators were either unsure of or indifferent to. For eradicators, time was of the essence, and aside from the problem of resistance, DDT remained an efficient and effective means of killing insects. Moreover, its long-term toxicity as a residual spray meant that eradicators only had to apply the chemical two times per year for effective control. So despite the real problems with resistance and the mounting evidence of ecological damage, DDT remained a powerful and popular chemical pesticide throughout the 1950s and 1960s.

Yet even for scientists, the way the chemical worked remained a mystery. A study reported in 1956, "The mode of action of DDT is not known." Another study published in 1959 concluded, "The mechanisms by which DDT and its analogues unstabilize nervous tissue is far from clear."[60] Even today, scientists remain inconclusive about how DDT works, although many theorize that it disrupts sodium channels within nerve cells, thus incapacitating insects.[61] What was clear in the minds of many eradicators was that the chemical worked.

Ecologists, however, remained skeptical of the "wonder chemical." The mounting evidence of fish and fowl dying from massive spray campaigns reaffirmed the dangers of DDT, even as eradicators attempted to mitigate these problems. But what was perhaps more problematic was DDT's ability to migrate. Unlike organophosphates like malathion and parathion, which were water soluble and broke down fairly rapidly once applied, DDT concentrated within fatty lipids of animal species as it moved up the food chain, and did not break down in water. Therefore, once applied, the chemical could not be contained within its local environments. And as DDT was being deployed for myriad reasons, near and far, the impact of the chemical had global consequences.

Throughout the United States, public health officials and agricultural experts dispersed DDT and other chlorinated hydrocarbons with reckless abandon. In Illinois, public health officials deployed army trucks and a B-25 bomber to spray DDT in an effort to "curb the rise of infantile paralysis," (polio) thought to be transmitted by flies.[62] Other campaigns dispatched trucks that dispersed a chemical fog, coating suburban neighborhoods with high concentrations of DDT. And farmers and housewives deployed the chemical to rid fields and homes of unwanted pests. Yet with each use, DDT was writing its own history. For example, by 1947 large areas of the Northwest were being sprayed with DDT in attempts to control the Douglas fir tussock moth.[63] In the eastern United States, the threat of a gypsy moth infestation launched similar spray programs, which lasted several years. In both cases, spray planes covered broad swaths of forested lands with a fine mist of DDT immersed in petroleum. And in both cases, DDT migrated from its targeted destination.

In the case of the tussock moth, for instance, the DDT sprayed over the Blue Mountains of northeastern Oregon most likely fell into the Grande Ronde River, a tributary of the Snake River, which itself is a tributary of the Columbia River. Once ensconced in the riverbed, the DDT molecule (and its metabolite, DDE) could simply ride the currents, making the 500-mile journey to the Pacific Ocean. There it might have encountered the seasonal migrations of the Pacific herring, a species which, while feeding on plankton or smaller fish species, may have ingested DDT. Once consumed, the DDT molecule enters the herring's bloodstream and is ultimately absorbed into its fat cells, where the chemical will remain. Since these schooling fish are an important source of food for seals, sea lions, whales, and aquatic birds, DDT was likely passed along during feeding. And as predatory animals consume animals carrying trace amounts of DDT within their bodies, these molecules tend to accumulate within the predatory animal at higher concentrations than in the original prey. This is a process biologists call biomagnification. And it is a process that occurs within all living animals. Over time, scientists began discovering trace amounts of the chemical in Arctic wildlife, even though the chemical had never been used in the region.[64]

Somewhat differently, the release of DDT over the Blue Mountains may have followed such a path, only this time it might have been consumed by young salmon ready to take their journey to the open ocean. No longer than five inches, the salmon fry travels to the Columbia River, making the first half of the migration that eventually brings it back to its place of birth. Unfortunately, the river's powerful currents lost speed as the river collided with the Bonneville Dam, a relatively new project completed in 1938 that was designed to bring the transformative power of electricity to the region. With their progress slowed by the dam, the salmon were vulnerable to predatory birds like cormorants, eagles, and egrets. Over time, the reproductive capacity of these predatory birds may have become compromised because high concentrations of DDT resulted in thinning eggshells and premature births. Because of its persistency and ability to migrate, DDT's long-term environmental impacts were both local and global. Yet very few understood the breadth and scope of DDT's migration and the ecological problems it engendered.

Eradication, however, superseded all of these concerns. Rather than define the world's malaria problem as part of a political, cultural, or ecological process, one shaped by a long history of human and nonhuman actors, eradicators understood the disease through an ever-narrowing lens of vector control. By funneling time, resources, and human energy into the single (and universal) strategy of eradication, public health experts demonstrated their belief in the power of human technologies. Indeed, because the American Century was based on the idea of remaking the world beyond U.S. borders, eradicators used DDT to recast diseased-plagued areas of the world into what they hoped would be vibrant capitalist economies. Belief in the chemical's ability to accomplish such a vast undertaking cemented DDT as an essential part of a much broader strategy aimed at promoting American interests throughout the world.

The ideology of eradication, shaped in large part by Cold War tensions, presupposed the supremacy of human technologies. And while many other Cold War technologies were greeted with similar enthusiasm—atomic energy, for one—DDT engendered an almost mystical fascination throughout much of the United States and the developed world, particularly because the solution to the historic problems of disease and famine seemed so close and, on the surface, easy.

Yet nature fought back, as most eradicators were aware at the time. Insect resistance and DDT migration revealed that nature was not a passive actor. Instead of retreating, however, eradicators simply renewed their faith in chemical technology. Whether on a national or global scale, postwar eradication required big plans. And it was the grandness that engendered enormous enthusiasm for the idea. Not unlike the process of urban renewal that Lewis Mumford so notably detested, which sought to eradicate "diseased" neighborhoods with the might of federal dollars, municipal housing authorities, and the wrecking ball, insect eradication was designed to cast out old, diseased environments in order to unleash the productive capacity of healthy people. It was this same philosophy, however, that undermined eradication efforts over time. The scale of eradication, whether in the United States or abroad, marginalized local differences. Although WHO records make it clear that public health officials paid attention to local conditions and were often confronted with environmental conditions that challenged their best efforts to eradicate mosquitoes, it is also clear that the solutions to complex problems were always found at the end of the spray gun.

Looking back at the evolution of eradication during the twentieth century, we can clearly see the impact DDT had on the eradication narrative.[65] Rather than seeing eradication as a process rooted in local environments as Gorgas, Le Prince, and Hardenburg did, postwar entomologists, public health officials, and agriculturalists marginalized broad-based approaches to insect control and universalized the eradication model. By developing a single solution to the multifaceted dimensions of the "insect problem," they reduced or essentialized nature as a universalized whole that human action, regardless of ecology or history, could change at will. And despite earlier efforts to investigate the ecologies of disease prior to and during the Second World War, the belief in DDT and other chemical pesticides made these ecological studies less important.

The age of wreckers and exterminators, as the nuclear age had been, was fraught with danger and possibilities. And at the heart of this conflict lay the scientific technologies that could either enrich or destroy the world. A. V. Hill remarked in his presidential speech before the British Society for the Advancement of Science in 1952, "It is true that scientific research has opened up the possibility of unprecedented good, or unlimited harm, for mankind but the use that is made of it depends in the end on the moral judgments of the whole community of men. . . . To help to guide its use is not a scientific dilemma, but the honourable and compelling duty of a good citizen."

Ten years later, Rachel Carson would embrace the role of the "good citizen." Her book *Silent Spring* would expose the logic of eradication, condemning those who believed they could control nature and opening up a new ideological terrain that placed humans squarely within the natural world—a terrain known as ecology.

5

GREEN REVOLUTIONS
IN CONFLICT
Debating *Silent Spring*, Food, and
Science during the Cold War

In spite of the apparent conscientious effort on the part of Miss Carson to be of
service to the public, her failure to discuss the benefits derived from insecticides
for insect control or to consider the consequences if they are not used
thereby permitting the public to weigh the advantages and
disadvantages of their use represents a serious omission.
—Edward F. Knipling, ARS director, 1962

In 1961, the United Nations launched an ambitious project to "promote social progress and . . . to employ international machinery for the advancement of the economic and social development of all peoples." Embarking on a "Decade of Development," the United Nations challenged member nations to use private and public funds to accelerate economic development in "less developed countries through industrialization, diversification and the development of a highly productive agricultural sector." This campaign was to be the beginning of a dramatic transformation of the Third World. It was a call to arms to help those who "needed" assistance, a call based on a shared belief in human decency, morality, and peace. It was also a means to integrate large parts of the world into a global economy. Solving problems of disease and malnutrition was seen as the key to reversing the economic plight of developing nations.[1]

Newly elected president John F. Kennedy shared many of these assumptions. In his inaugural address, Kennedy spoke of "casting off the chains of poverty," proclaiming to "those peoples in the huts and villages across the globe struggling to break the bonds of mass misery" that the United States pledged its "best efforts to help them help themselves . . . because it is right." During his short-lived presidency, Kennedy established the Peace Corps, which sent thousands of eager young Americans across the globe as humanitarian and technological ambassadors, and spent billions of dollars forging ties with Latin America through the Alliance for Progress program. In 1961, Kennedy signed the Foreign Assistance Act, requiring the federal government to establish a separate agency for economic aid, thus creating the United States Agency for International Development (USAID). Collectively, these institutions would play a critical role in applying American technology to reshape foreign lands in the decade to follow.[2]

Kennedy's commitment to foreign aid reflected the realities of a world in flux. The Cold War and the independence movements in Asia and Africa framed many of the decisions made in the name of humanitarianism and economic development. Kennedy funneled aid to countries considered vital to U.S. interests, specifically India, Pakistan, and Turkey, all of which received enormous amounts of financial assistance and food in the U.S. effort to thwart possible communist influence. As one historian notes, "Kennedy viewed the strong and rapidly developing India as a counterbalance to the influence of the People's Republic of China" and believed friendship with India "would demonstrate Washington's sympathy for third world nationalism."[3]

As earlier postwar aid programs had been, the Decade of Development was predicated on the curative power of modern technologies. Development experts believed that the transfer of technological knowledge could change the economic plight of poor nations. Arguably, the most comprehensive development project was the Green Revolution, an ambitious program to transform rural lands into highly productive agricultural "factories." Based on the development model created by the Rockefeller Foundation in Mexico, the Green Revolution required an enormous amount of technical inputs to increase agricultural yields. Hybrid seeds, chemical fertilizers and pesticides, and tractors were used to awaken the agricultural and economic dormancy of developing nations. Moreover, loan guarantees by the World Bank and the International Monetary Fund encouraged and required developing countries to incur much of the cost of the Green Revolution as well. By investing in modern technologies, many experts believed, rural economies could throw off the shackles of poverty, malnutrition, and perceived "backwardness" and become productive members of a global economy. Over the course of the decade, the political and economic forces that reshaped farmlands in much of the world changed how people lived. However, as many critics of the

Green Revolution contend, the change was often to the detriment of these rural communities.[4] Within the span of twenty years, and for myriad reasons, the promise of development would give way to years of unfulfilled promises and disappointment.

Within the United States, however, another Green Revolution was about to unfold. In 1962, Rachel Carson, a highly successful but relatively obscure nature writer, released her third book, *Silent Spring*, a book that, one historian proclaimed, "ranks as one of the most important books ever published by an American writer."[5] Selected by *Time* magazine as one of the top hundred books of the twentieth century, *Silent Spring* transformed the debate over DDT, leading many to credit Carson with starting the American environmental movement. While depicting Carson as the progenitor of American environmentalism largely ignores the longer and more complicated history of the movement, Carson was certainly an important voice in the development of a postwar environmental movement.

Silent Spring shifted the debate about DDT. Although scientists had published numerous articles on the risks and benefits of the "miracle" pesticide since 1943, much of it remained within the community of entomological and ecological science. *Silent Spring* allowed readers into a world that had been closed to nonscientists and made complex scientific concepts accessible to a general audience. By unlocking the mysteries of the chemical sciences, Carson exposed the failure of industrial science to contend with a series of biological and ecological issues that affected human and nonhuman life. Her popular discourse on ecology encouraged readers to examine the various connections between the human and nonhuman world, shaping a social movement that merged ecology and grassroots activism, a movement one scholar called "the Green Revolution."[6]

The Decade of Development launched a Green Revolution on a global scale based on the transformative promise of technology. Carson's *Silent Spring* engendered a Green Revolution within the United States defined by the science of ecology. The disparity between the two Green Revolutions—one international and technological, one national and ecological—underscores the political contradictions that emerged during the early 1960s. The international aid community saw chemical pesticides as a tool necessary for building strong bodies and sound economies, while Rachel Carson posited otherwise. She argued that because they were persistent and pervasive, DDT and other hydrocarbon pesticides endangered the human and nonhuman worlds. Each side in this revolution was caught between conflicting ideas about the promise and dangers of chemical pesticides. The Cold War only intensified this debate. The global struggle against communism placed the domestic DDT debate squarely within a divisive political conflict that drew clear distinctions between the hungry and the well fed, the healthy and the infirm, and the free and the oppressed. For many, DDT was not simply a chemical compound that eliminated insect pests, but a critical element for the defense of the nation. Criticisms of the pesticide, like those posited by Carson, therefore, were seen by some as threatening the nation's well-being at home and its interests abroad. The debate that emerged after the publication of *Silent Spring* was not simply about the meaning of nature and chemicals in the modern world; it also reflected the global impact of DDT and the Cold War.

SECTIONS OF *SILENT SPRING* first appeared in serial form in the *New Yorker* during the summer of 1962. By August, editorials, news stories, and letters to the editor appeared in *Chemical Week, Chemical and Engineering News, Facts for Farmers, National Audubon Magazine, Newsweek,* and *Rural Virginian.* The *New Yorker* received well over 200 letters, the vast majority of them favorable to Carson's position. By September, most major newspapers had published stories or editorials about Carson's work. *Silent Spring* became an instant sensation by the time it appeared on bookshelves later that September. It quickly rose to the top of the *New York Times* bestseller list, where it remained for the next thirty-one weeks. Rachel Carson had touched a nerve with the American public, a large portion of which had come to realize the environmental consequences of postwar prosperity.[7]

Born in 1907 in the rural hamlet of Springdale, Pennsylvania, north of Pittsburgh, Carson grew up enthralled with nature and books—two interests that would be consistent in her life. As a young child, she wrote articles on the wonders of the natural world, publishing her first story, "A Battle in the Clouds," at the age of eleven. In 1926, Carson entered Pennsylvania College for Women (now Chatham College) in Pittsburgh, graduating magna cum laude with a degree in science in 1929. Three years later, she completed a master's degree in zoology at Johns Hopkins University. Graduating at the height of the Great Depression, Carson found employment as a staff writer within the Bureau of Fisheries. Her knowledge of scientific research and her talent as a writer were readily apparent, and she became head of the bureau's publications division in 1949. While at the Bureau of Fisheries, Carson published a series of articles on ocean life that was widely regarded for its scientific and literary strength. In 1941, she completed her first book, *Under the Sea Wind,* which received positive reviews for its mixture of poetry and science. However, it was the publication of Carson's second book, *The Sea Around Us*

(1951), that earned her critical acclaim, including the National Book Award, the Burroughs Medal (commemorating excellence in nature writing), and recognition from the *New York Times* for an outstanding book of 1951. Because of this recognition, Carson had the financial independence to leave her post at the Bureau of Fisheries in order to write full time. The publication of *Silent Spring* made her one of the most important writers of her generation.

Simply put, *Silent Spring* changed the DDT story. But *Silent Spring* was not only an exposé of the chronic dangers of DDT and other chemical pesticides; it was an indictment of the logic of eradication and the way Americans understood the environment and their place in it. Carson's ability to unlock the mysteries of science enabled her readers to grasp the complexities of the modern world. Lying just underneath the shiny veneer of American postwar prosperity, Carson suggested, was a multitude of chemical interactions that endangered the environment, people, and, by extension, the future of the nation. The glossy exterior of postwar America—the tree-lined suburbs, the automobiles, and the trappings of a consumer republic—created a false sense of security that was undermined by dangers no one could see. "For the first time in the history of the world," Carson wrote, "every human being is now subjected to contact with dangerous chemicals, from the moment of conception until death." She convincingly exposed the dangers of the chemical infrastructure sustaining much of the American postwar economic boom. DuPont's advertising slogan "Better living through chemistry" never looked more disquieting, as Carson held that chemistry had devastating consequences, not only for her contemporaries, but for future generations.[8]

Moreover, these threats came from everywhere: housewives who sprayed their homes with the latest chemical concoction to reduce insects, men who applied an assortment of chemicals to keep their lawns pristine, farmers who were required to spray their fields more frequently with each passing year because of the reduced effectiveness of pesticides, and public health officials who administered the deployment of millions of pounds of chemicals throughout the nation in a near-religious quest to eradicate insect pests. Through it all, these "elixirs of death," as Carson so provocatively called them, indiscriminately endangered the lives of all people with a "chemical death rain" designed to kill insects. As one looked beyond the sanitized landscapes of modern America and the "smooth superhighway on which we progress with great speed," Carson warned, there lay "disaster."[9]

For Carson, the pesticide problem was inescapable. And this inescapability was perhaps the greatest danger. "The central problem of our age," Carson maintained, "has therefore become the contamination of man's *total* environment with such substances of incredible potential harm."[10]

To illustrate the concept of "total environment," Carson opened her book with a parable about the consequences of chemical technologies. The chapter "A Fable for Tomorrow" depicts a town laid waste by a "strange blight" causing "a shadow of death" to appear wherever one looks. Imagined to be somewhere deep "in the heart of America," this fictional landscape was "once so attractive," but is now "lined with browned and withered vegetation as though swept by fire." The town's stillness is matched only by its silence; it is a landscape "deserted by all living things." Images of decay, destruction, and death are made even more excruciating by the suggestion that human action caused them. "No witchcraft," Carson wrote, "no enemy action had silenced the rebirth of new life in this stricken world. The people had done it themselves." This is a world created by the reckless misuse of chemical pesticides. It is a place Carson feared might exist if Americans did not change their understanding of the natural world and their place in it.

Perhaps more troubling was the speed at which real-world change occurred. In ecological time, it was a mere blink of an eye. "Only within the moment of time represented by the present century has one species—man—acquired significant power to alter the nature of his world," Carson asserted. Within the span of a few years, the radical transformation of nature by humans wielding an assortment of chemical technologies raised serious concerns about the future. If spray programs continued unabated, what would happen to the natural world? Would the sounds of spring remain silent? Or, if left to their own devices, would insects outlive humans?

To Carson, the ecological hazards of pesticides were clear. But what was more disturbing than the omnipresent threat of chemicals in the environment was the logic that encouraged their use. Eradication, for instance, served to illustrate the deep disconnect that humans had with the natural world. Using modern technologies to fundamentally transform nature was antithetical to Carson's ecological worldview. Not only did proponents of eradication fail to grasp the complexities and interdependencies of nature, but they presupposed the moral authority that human beings had in selectively deciding which species to eradicate and which to save.

To some, the mosquito and fire ant eradication campaigns demonstrated the futility of trying to control nature. Certainly the dead bodies of countless insects were testament to the power of chemical pesticides, but the biological adaptability of mosquitoes and fire ants—and other insects, for that matter—undermined the power of these chemical technologies in what Carson called a "flareback." To those who considered humans the masters of the natural world, Carson offered a stern warning: "The chemical war is never won . . . [and] all of life is caught in its violent crossfire."[11]

This "violent crossfire," Carson argued, had severe consequences for both the human and the nonhuman worlds. Her description of how chemical pesticides affected the human body opened up a discursive path to address the links between environment and health. A border between nature and the human body did not exist; rather, the two were intrinsically linked in a biological web that was as timeless as it was natural. According to historian Linda Nash, "Carson put forth a popular discourse of ecological health that would challenge the models promulgated by contemporary medical experts."[12] Cancer, chromosome abnormalities, and other chronic ailments called attention to the impact long-term exposure to chemical pesticides had on the human body. It was DDT's chronic, rather than acute, danger, Carson asserted, that posed the greatest threat. The threat, as Carson saw it, was not the single application of or isolated contact with persistent pesticides—although some, like aldrin and dieldrin, were dangerous in those cases—but the long-term effects of small amounts of exposure, whether from breathing, eating, or contact with skin. Rather than creating healthy bodies, Carson argued, chemical pesticides poisoned them.

The central question posed in *Silent Spring*, however, was not whether pesticides were good or evil, but how much and in what context. In perhaps the most overlooked passage of *Silent Spring*, Carson posited that chemical pesticides, if applied judicially and with great care, had enormous potential. She wrote, "It is not my contention that chemical insecticides must never be used. I do contend that we have put poisonous and biologically potent chemicals indiscriminately into the hands of persons largely or wholly ignorant of their potentials for harm."[13] Evidently, in this case, the ignorant users were not the Mexican farmers described by Elvin Stakman, but the economic entomologists and development experts who willfully neglected the ecological and biological hazards of pesticides. While it is clear that Carson's book was an attack on the widespread, and oftentimes uncritical, use of chemical pesticides, Carson did not discount chemical technologies outright. Rather, she understood that chemical pesticides could be used in conjunction with biological and cultural insect control methods. Carson was an early advocate of what would come to be called integrated pest management, a multitiered approach to the "pest problem" that used chemical and biological methods to control insect pests, not eradicate them.

In April 1963, CBS television aired the first of two reports on *Silent Spring*, "CBS Reports: The Silent Spring of Rachel Carson." This highly anticipated news broadcast attempted to illuminate the many sides of the pesticide debate. Moderated by television journalist Eric Sevareid, the program included interviews with Carson, USDA secretary Orville Freeman, Food and Drug Administration head George Larrick, and American Cyanamid chemist and industry spokesman Dr. Robert White-Stevens, as well as representatives from the Interior Department's Fish and Wildlife Service, the U.S. Public Health Service, and the president's science advisory committee, which would soon release a report supporting Carson's contention of pesticide abuse. Yet what emerged over the airing of the two episodes effectively narrowed a complex environmental and human health issue into a simplistic, two-dimensional debate. Blunt and often verbose, White-Stevens argued vociferously against Carson's ecological position, for all intents and purposes eliminating any common ground between the two positions. "If man were to faithfully follow the teachings of Miss Carson," White-Stevens claimed, "we would return to the Dark Age, and the insects and diseases and vermin would once again inherit the earth." While the program was designed to bring diverse opinions together, the effect was to stir controversy rather than inform the audience. According to media scholar Priscilla Coit Murphy, who examined coverage of Carson and *Silent Spring*, the media, "to avoid the appearance of appointing themselves judges, . . . often resorted to a thorough airing of both sides, thereby making themselves not just sources of information but also catalysts of the controversy."[14]

The program generated an enormous public response, which eventually forced Congress to investigate the scientific evidence presented in *Silent Spring*.[15] Connecticut senator Abraham Ribicoff assembled a committee to study the "pesticide issue." The main purpose of the committee hearing was "to examine the role of the Federal Government as it deals with one of the great problems of our time: man's contamination of his environment." The proceedings were also an attempt to comprehend the control and jurisdictional mechanisms in place at the national level, and to determine if any further legislative action was necessary.[16] At issue were the jurisdictional responsibilities (or might) of the various government agencies that used, traded, regulated, or protected citizens from pesticides, agencies including the U.S. Department of Agriculture, the Food and Drug Administration, the Department of the Interior, and the Department of Health, Education, and Welfare. Prior to *Silent Spring*, most pesticide testing assessed a chemical's acute toxicity to humans, not its chronic effect. The burden of proof, moreover, was not placed on the chemical industry to determine toxicity, but was assumed by government officials at the Department of Agriculture, who were themselves interested in promoting agricultural pesticides. The star witness of the hearing was Rachel Carson, who was welcomed to the committee hearing by Senator Ribicoff, who said, "You're the lady who started it all." After giving a prepared statement explaining the ecological problems associated with excessive pesticide use, Carson responded to a question concerning the indiscriminate use of pesticides, stating, "I think that instead of automatically reaching for the spray gun or calling the spray planes, we must consider the whole problem."[17] Unfortunately, the hearing would be one of her last public appearances. Carson died one year later from breast cancer, but her legacy would shape the politics of the American environmental movement for the decades to come. Meanwhile, the Ribicoff committee failed to generate legislative movement in the halls of Congress. It did, however, bring national attention to the pesticide problem.

Carson's death left a profound void in the debate over DDT. In Carson's absence, critics, both domestic and international, would comment on, critique, celebrate, and even admonish her book. Moreover, the debate that emerged after the publication of Silent Spring quickly hardened into two opposing camps: one that valued the effectiveness, the low cost, and the low toxicity of DDT, and one that viewed the chemical as an environmental contaminant on a global scale. This division did not occur by happenstance, nor was it exclusively an effect of the media. Rather, the division resulted from the role of U.S. foreign aid during the Cold War, the politicization of science, and, as I will discuss in the next chapter, the legal methods employed by environmental activists to stop the use of DDT.

Largely absent from the political controversies about DDT, however, was Carson's claim for a rational approach to chemical technologies that did not border on fanaticism. Although many of her contemporary critics painted her as a fanatical defender of nature, she did not reject chemical technologies, nor did she refute the science behind them. Instead, she understood that chemical technologies had a place in the world, but that world required humans to rethink their connection to it. For her, ecology meant reconnecting the human world with the natural. The long-term health of the human world, she argued, depended on forging a new symbiotic relationship with nature. How one went about this, however, remained a vexing question.

Others continued to see the natural world differently. Not surprisingly, the chemical industry quickly drew sharp distinctions between the use of chemical pesticides and Carson's ecological approach to chemical controls.

Tying the pesticides debate to the development project, domestic chemical producers quickly challenged Carson's assertions with arguments using the divisiveness of Cold War politics. Chemical pesticides were promoted as a critical element in the war against communism because they provided "protection" against the insect menace. The competing Green Revolutions, therefore, were rendered not in shades of grey, but the starkness of black and white.

Indeed, malaria eradication and agricultural development projects cast a long shadow over the contentious politics that emerged after the publication of *Silent Spring*. Moreover, mounting concern over population growth shifted the focus away from malaria eradication programs toward agriculture. For many development experts, the Green Revolution offered the best chance to solve the alarming challenge of feeding millions of undernourished people. Agricultural experts maintained that marginal lands could be transformed into productive lands through the strategic deployment of agricultural technologies. To meet these challenges, agricultural experts and the development community built a series of new agricultural research stations during the 1960s based on the Rockefeller Foundation's project in Mexico. Mexico became the site of a new wheat research station, and a completely new project was launched in the Philippines to developing high-yielding strains of rice.

In Mexico, for example, the seeds of the Green Revolution began to blossom and proliferate. Though the Rockefeller Foundation dismantled its Mexican Agricultural Project (MAP) in 1961, the foundation continued to support agricultural development under the rubric of the newly formed International Corn and Wheat Research Institute (ICWRI). Due to bureaucratic troubles, the ICWRI was reorganized in 1966, operating under the name International Maize and Wheat Improvement Center (Centro Internacional de Mejoramiento de Maíz y Trigo, or CIMMYT). Funded by grants from the Rockefeller Foundation, the Ford Foundation, USAID, and the United Nations Development Program, CIMMYT led the charge against hunger on a global scale.[18] Two MAP stalwarts, Edwin Wellhausen and Norman Borlaug, headed the new project. Wellhausen served as the project's first general director, while Borlaug spearheaded the wheat program. Yet it was Borlaug who would become the public face of the development project and the Green Revolution. One architect of the U.S. Food for Peace program, Don Paarlberg, referred to Borlaug simply as the "Hunger Fighter."[19]

Ever since he began working in Mexico, Borlaug promoted agricultural science as a way to reverse the debilitating impact of famine. His dwarf wheat variety not only revolutionized the production of wheat in Mexico, but it came to symbolize the power of agricultural science and technological proficiency. Sonora 63 and Sonora 64, the hybrid seeds that changed the face of Mexican agronomy, represented the power and promise of postwar technology. For his part, Borlaug became an outspoken advocate for the technological revolution he engendered. "Seeds," he once claimed, "are the catalyst. You have to support them with a total package of modern technology—and with strategy, with psychology, and a firm, guaranteed economic policy from the government."[20] For Borlaug, the total package, including, of course, chemical fertilizers and pesticides, represented a triumph of human ingenuity, political will, and modernity.

In 1963, Borlaug led an agricultural envoy to India to determine how best to increase crop yields in a country facing severe grain shortages. Dissatisfied with Indian efforts to use modern hybrids, Borlaug expressed dismay at the experimental plots, which were underfertilized and had "miserable weed control." After his trip, Borlaug oversaw the shipment of over 500 kilograms of his Mexican hybrid wheat seeds to India. Shipments of seeds would continue to flow into India throughout the 1960s as droughts and concerns over overpopulation heightened fears of widespread famine. "By 1965," Borlaug later remarked, India had "decided to import 250 tons of seed of the Mexican dwarfs," although the shipment almost did not arrive in time for the fall planting because it was held up in the port of Los Angeles as a result of the Watts Riots.

Responding to the changing global problems, the Rockefeller Foundation created two projects to tackle the joint problems of population and famine, projects named the "Problems of Population" and "Toward the Conquest of Hunger," respectively. In the early 1960s, the Ford Foundation also increased its presence worldwide as the issues of food production and population growth became more pronounced. Together with the Rockefeller Foundation, the Ford Foundation financed the International Rice Research Institute (IRRI) in 1960, which was designed to develop high-yielding varieties of rice for global distribution.[21]

These international scientific institutions represented the high point of postwar technological modernity. The IRRI, for instance, expressed the modernist ideal through innovative scientific research and by stylistic design, evident in the brilliant simplicity of the institute's shimmering white buildings set against the tropical landscape of the Philippines.[22] American foundations remained a central funding source for the IRRI's transformative project.

Within a few short years, however, multinational corporations, many of which would privatize agricultural development and secure propriety rights for seeds, fertilizers, and pesticides, would overshadow these once-prominent research centers. This shift was neither entirely complete nor swift. Throughout its agricultural adventures, for instance, the Rockefeller Foundation relied on the cooperation of U.S. corporations to support its research efforts. Producers of pesticide and fertilizers were critically important to the foundation's agricultural work. The connections between agricultural research funded by foundations and other aid organizations and agribusiness strengthened over time. By the end of the decade, U.S. multinational corporations began to take a more active role in promoting, as well as defending, their vision of industrial agriculture.

In the early 1960s, the Kennedy administration increased its foreign food and technical aid as well. Relying on the "social science" work of senior foreign policy adviser Walt W. Rostow, whose influential book *The Stages of Economic Growth: A Non-Communist Manifesto* (1960) shaped much of Kennedy's thinking about aid, the White House strategically deployed federal dollars to alleviate suffering overseas and protect U.S. interests abroad. From 1960 to 1965, U.S. investment in foreign aid increased from $2.8 billion to $3.6 billion. Rostow's modernization theory presupposed that human development was a linear process in which a nation's economic growth could be charted along an axis. Given the right factors, developing nations could "take off" to become nations of "high Mass Consumption," which, according to Rostow, was the highest order of economic development. "Rostovian theory," political scientist Kimber Charles Pearce wrote, "stressed the commonalities in the evolution of developing nations, but masked significant social, political, and economic differences among nations."[23] Indeed, Kennedy's foreign aid initiatives underscored the belief in universal human dignity, yet their application resulted in a highly authoritarian, almost monolithic approach to agricultural development and public health, as did the aid initiatives of the Johnson administration.[24]

Ecological difference did not matter to many development experts. The universal approach to the insect problem called attention to the limits of the development project, which presupposed the commonality of the human and nonhuman worlds. Nature, regardless of geography or the cultural legacy of land use, could be transformed to become productive and modern. In the eyes of development experts, insects and poor soil composition were merely impediments to the modernizing impulses of the development project. Chemical fertilizers and pesticides were tools for bringing about the modern world, regardless of the ecological damage they might cause.

While largely a book about pesticides, *Silent Spring* was, in many ways, a critique of postwar modernity. Carson challenged one of the underlying assumptions of postwar America, the belief that science and technology could improve the health and economic fortunes of the nation. In fact, she rejected this premise outright. Not only did science and technology unleash an assortment of deadly chemicals indiscriminately throughout the natural world, but the logic behind the spray gun undermined the ecological connections humans had with the natural world. "The 'control of nature,'" Carson wrote in the last paragraph of Silent Spring, "is a phrase conceived in arrogance. . . . It is our alarming misfortune that . . . science has armed itself with the most modern and terrible weapons, and that in turning them against the insects it has also turned them against the earth."[25]

Amid the escalating tensions of the Cold War, and in the context of the prevailing belief in agricultural development as a viable political and economic solution to the world's problems, the role of science and the scientist shaped the political discourse over Silent Spring. Even before Silent Spring hit bookshelves, U.S. chemical producers were drawing distinctions between the sides of the pesticide debate. Chemical producers saw themselves as allies to American cold warriors and the development project. After the book's publication, however, domestic chemical producers faced what one industry official called "a public relations problem."[26]

The chemical industry tried to mitigate this problem in two ways. First, it questioned Carson's scientific expertise, which was difficult given her professional credentials, raising the issue of professional authority through a strongly gendered filter. The industry's attacks on Carson's expertise emphasized the concept of scientific militarism and presupposed the inherent masculinity of Cold War technologies, while writing off ecological science as simply a feminine pursuit. The chemical industry held up Carson as the most notable example of an ecologist out of step with the nation's interests. Second, the chemical industry strategically invoked the international successes of DDT as a way of rejecting the claim that chemical pesticides were dangerous at home. These two industry responses opened up a series of questions about environmental health, international development, and the Cold War. They also revealed a debate brewing within the scientific community over professional authority, knowledge production, and the role of the scientist within the public realm.

Silent Spring exposed one of the underlying tensions of Cold War science, namely, who spoke for science. In this regard, Carson instigated a debate within the profession about knowledge production and professional authority that pitted the "soft" science of ecology against the "hard" science of chemistry. Set within a global conflict, these battles took on greater significance. The Cold War demanded that new scientific institutions keep pace with the enemy. By 1960, nearly all aspects of scientific discovery were shaped by the global conflict.[27] Eisenhower's warning about the military-industrial complex reshaped the links between scientific institutions—both public and private—and the state. Weapons production, national security, and the heightened concern over the communist threat transformed scientific institutions into a strategic arm of U.S. military might. Vannevar Bush's dream—though some would call it a nightmare—seemed to have come to fruition.[28] Domestic consumption of pesticides and other chemical components during the Cold War was seen as the continuation of the war against the oppressive forces of communism.[29]

Within this heightened military context, the gendered divisions of the natural world were drawn much sharper as the masculine connotations of war-making dominated the public discourse.[30] The "hard" chemical sciences, which advocated "eradication," "annihilation," and "all out war" against insects, were in direct opposition to Carson's ecological thinking, which was described with terms like "nature loving," "innocent," and "harmonious." Indeed, Carson's critics charged her with being "emotional," "hysterical," and an "alarmist" whose methods were "un-scientific." *Time* magazine claimed *Silent Spring* was an "emotional and inaccurate outburst."[31]

Others lampooned her literary style, calling it "un-scientific," yet "graceful." It was a book, they claimed, "written in glowing scarlet."[32] Edwin Diamond, a critic who had once worked with Carson in the early research stages of *Silent Spring*, wrote in his review for the *Saturday Evening Post*, "Her arguments were more emotional than accurate," yet the book "has a high expository gloss."[33] Pincus Rothberg, the president of Montrose Chemical Corporation, one of the major manufacturers of DDT in the United States, claimed that Carson wrote not "as a scientist but rather as a fanatic defender of the cult of the balance of nature."[34] Fredrick Stare of the Department of Nutrition at the Harvard School of Public Health claimed, "Her book is an emotional picture. . . . Carson writes with passion and with beauty but with very little scientific detachment. Dispassionate scientific evidence and passionate propaganda are two buckets of water that simply can't be carried on one person's shoulders. . . . Miss Carson flounders as a scientist in this book." Even UNICEF's head of applied nutrition, L. J. Teply, commented, "I too was disappointed that Miss Carson was not more objective."[35]

In the cold, calculating world of Cold War science, this gendered critique of Carson not only discounted her role as a woman of science, but also the scientific principles of ecology she espoused. If one looks at the reviews for Carson's earlier books, however, this gendered language becomes clearer within the context of industrial science. Carson won immediate, and almost universal, praise for *The Sea Around Us*, earning her numerous awards. Yet initial reaction to the book questioned Carson's scientific acumen because she was a woman. Carson's biographer, Linda Lear, has argued, "Many male readers, and certainly the scientific community, were reluctant to admit that a woman could deal with a scientific subject of such scope and complexity."[36] Despite this initial reaction, as well as the curiosity of reviewers "to know what this woman looks like," Carson's position as a nature writer was never subsequently questioned. *New York Times* science editor Jonathan Leonard wrote, "When poets write about the sea their errors annoy scientists. When scientists write about the sea their bleak and technical jargon paralyzes poets. Yet neither scientists nor poets should object to 'The Sea Around Us.'"[37] Likewise, when *The Edge of the Sea* was published, Carson earned praise for her "factual poetry," as one reviewer noted.[38] *Time* magazine wrote that once again "Carson has shown her remarkable talent for catching the life breath of science on the still glass of poetry."[39] Within the realm of nature and the ebbs and flows of the ocean currents, poetry had its place and women were welcome.

By questioning *Silent Spring*'s literary convention, the chemical industry did more than simply critique Carson's scientific ability. Rather, the industry directly challenged the scientific legitimacy of ecology. Gender served as a means through which the technological sciences maintained their authority as the true holders of scientific knowledge. Moreover, the supposed softness of ecology ran counter to the geopolitical realities of the Cold War. H. Patricia Hynes's gendered analysis of *Silent Spring* claimed, "The patriarchs of agricultural politics and the chemical industry built their defense of chemical agriculture on the metaphor of war, with insects as enemies, chemicals as weapons, and themselves as heroes."[40]

As it happened, after Rachel Carson's *New Yorker* articles appeared, the first public statement by the chemical industry appeared in *Chemical and Engineering News*, on July 2, 1962. Ironically, the headline read "Pesticide Sales Pick Up." After painting an optimistic picture of chemical sales for 1962, the article foreshadowed potential problems ahead. Quoting the director of New Jersey's Department of Agriculture, F. A. Soraci, who was spearheading the state's gypsy moth aerial spraying program, reported, "In any large scale pest control program in this area, we are immediately confronted with the objection of a vociferous, misinformed group of nature-balancing, organic-gardening, bird-loving, unreasonable citizenry that has not been convinced of the important place of agricultural chemicals in our economy."[41] The quote seemed prophetic since *Chemical and Engineering News* and other trade magazines would spend the remainder of the year and much of the next decade trying to inform this unreasonable citizenry.

This critique did not go unnoticed, however. In a letter to the editors of *Chemical and Engineering News*, Dr. Paul Kronick countered by writing, "Your condescending reference to the 'claims' and 'adverse publicity' of 'conservationists,' [and] your quotation of Mr. Soraci's malignant characterization . . . of New Jersey's 'citizenry' completely [ignore] the fact that these 'nature-balancers' include a large number of biologists and zoologists whose credentials are impeccable in scientific circles." Kronick concluded his remarks with a pointed question that shed light on the tensions within the scientific community: "Are such credentials to be unacceptable to chemists?"[42]

The introduction of ecology into the public discourse over environmental health unequivocally challenged the scientific authority of those invested in the development project and industrial science. Moreover, and perhaps more critically, since the technological sciences undergirded the military apparatus of the country, serving a pivotal, if not primary, role during the Cold War, ecology seemed to have the potential, however unlikely, to undermine the safety of the United States.

So at a time when the authority derived from technological know-how went largely unquestioned, leading historian Elaine Tyler May to suggest that "when the expert spoke, postwar Americans listened," the popularity of *Silent Spring* and the science it espoused engendered an enormous amount of concern from those invested in technological modernity. For many, the book endangered not only the chemical industry, but the development project and the nation as well. As historian Michael Egan has suggested, "Ecology had become the subversive science."[43]

Ecology, while subversive in the sense that it addressed problems that had been ignored by other scientific disciplines, was not a novel concept. The word "ecology" is derived from the Greek *oikos*, meaning "household." Early usage of the word, however, conveyed a commitment to managing natural resources by "a community or a state for orderly production."[44] Over time, the meaning of the word changed, but ecology, as a consciously interdisciplinary endeavor, remained bound within a tight-knit community of scientists. However, the popularity of *Silent Spring* broadened the strict scientific usage of the term. Carson's ecological perspective provided a language to explain and articulate the sterile landscapes of Cold War America. So in a world shaped by the master plans—Levittown, Robert Moses's New York, and the USDA eradication campaign—popular ecology encouraged Americans to rethink the world around them. In their own right, popular American writers like William Whyte, Jane Jacobs, Betty Friedan, and John Kenneth Galbraith offered provocative analyses of the unrealized promises of postwar America. In one way or another, they wrote about isolation, arguing that the bureaucratization of the workplace, the transformation of the urban environment, suburbanization, and affluence created spaces that disconnected humans from each other and from the natural world.[45]

Ecology became the rallying cry of the modern American environmental movement, tying together concepts of environmental protection and ecological science that were not necessarily, or historically, connected. Therefore, ecology functioned within two realms: the scientific, where the study of interconnections exposed the shortsightedness of the chemical sciences, and the popular, where ecological ideas sought to reconnect humans to the natural world while at the same time encouraging people to question the authority of the expert.

Perhaps nothing raised ire within the scientific community more than Carson's expository choices and literary brilliance. Carson's short opening chapter, "A Fable for Tomorrow," generated the most disdain, largely because it was a fictional account masked as a possible future reality. Twice the chemical industry responded with parodies of the chapter, recasting the fable to shed light on the inability of ecological science to solve the problems of the modern world.

The first industry parody appeared on the editorial page of *Chemical Week* in August 1962. Written by the American Cyanamid director of research, Thomas H. Jukes, one of Carson's most vociferous critics, the editorial, "A Town in Harmony," depicted the dangers of living in harmony with nature. "The harmony of all life in this idyllic town," Jukes wrote, "followed a biological balance in Nature, a balance which man had not yet learned to disturb by drastic intervention on his own behalf."[46] Failure to shift the balance of nature led to a life of "scarcity and pestilence," in which "malaria," "whooping cough," "typhus," and "famine" weakened the human body and ultimately destroyed the town. Obviously, Jukes considered nature an inhospitable and dangerous place, which produced life-threatening conditions far worse than any problems chemical pesticides may have caused.

Yet he saved his most direct criticism for Carson herself. "The women," he wrote, "who, a century later, might have been writers of science fiction horror stories [i.e., *Silent Spring*] were too busy . . . squashing black beetles;

beating the clothes moths out of the winter woolens; scraping the mold from the fatback pork; and wondering if they could afford the luxury of a chicken for their Sunday dinner."[47] Jukes's commentary generated much criticism. In the weeks following Jukes's attempt at satire, *Chemical Week* published a series of responses, including one from a self-described "true scientist." "I had not expected something this vicious from you," the angry reader proclaimed. To some, the parody was "propaganda worthy of communism." Rather than respond directly to the charges, the editors of *Chemical Week* countered by claiming that Dr. Jukes was a "true scientist"—indeed he was—and to ensure that readers understood his professional credentials, the magazine listed his lengthy employment record and numerous awards.[48]

In October 1962, Monsanto Chemical Corporation published "The Desolate Year" in its monthly *Monsanto Magazine*. This too was a satirical piece that parodied Carson's literary style, depicting a world without pesticides where "food and fur animals weren't the only ones that died to the hum of the insects . . . man too sickened, and he died." As in Jukes's parody, nature was a menace to be controlled. Yet, in this case, Mother Nature not only had the power to destroy, but she had the power to emasculate technological enthusiasts. "The Desolate Year" described "hard-pressed men of the U.S. Department of Agriculture, besieged with pleas for help, [who] could only issue advisories to rake and burn . . . [and] Inspectors for the Food and Drug Administration . . . [who] were stumped for an answer." And a New York housewife whose apartment was crawling with ticks could not turn to pesticides and the manly exterminator. Monsanto asked, "What could she do? What *could* she do?"[49] Without the heroic efforts of the chemical industry and their weapons of war, Mother Nature would sway the balance in her favor, leaving once well-armed men helpless in her wake.

In constructing its assault against Carson and like-minded environmentalists, Monsanto integrated the ideas and imagery of "the foreign invader" into its defense of chemical pesticide use at home and abroad. "The Desolate Year," for example, used the image of the ruination of the United States in a world without pesticides. It pointed to the potential health and agricultural destruction that was at the nation's doorstep if pesticides were not available to the farmer or public health official. More importantly, however, one can also see references to the looming threat of some insidious other lurking beyond U.S. borders, waiting for the opportunity to strike: "Genus by genus, species by species, sub-species by innumerable sub-species, the insects emerged. Creeping and flying and crawling into the open, beginning in the southern tier of states and progressing northward."[50] Agricultural experts and members of the chemical industry would certainly have been familiar with this depiction in light of the recent "fire ant wars" that attempted to protect the nation from a foreign invader.

The threat of an outside invader, along with the desire to protect national interests, was central and consistent in reactions to *Silent Spring*. In 1962, the Pacific branch of the Entomological Society of America contended at its forty-sixth annual meeting that "the trouble is [that Carson] overlooked something pretty important: If we don't get them, they'll get us. Insects, if left to their own devices, are a bigger threat to our way of life than . . . communism."[51] The chemical industry adopted terms like "eradication," "menace," and "horde" to connect the insect problem with the ideological construction of anticommunism in the United States.[52] Indeed, bugs and communists were seen by many as dangers to the future prosperity of the nation.

Yet the struggle against communism was not simply fought on the home front. Rather, it was an international conflict, fought with guns and butter, to gain influence and elicit support from around the world, including support from many newly independent nations in Africa and Asia. Insect control and the politics of the Cold War were never far apart. So while the protection of the home front served as a powerful metaphor in the defense of chemical pesticides, the conflict beyond U.S. borders was equally important. American support for international development strengthened the claim that pesticides were of critical importance in the war against famine, disease, and communists.

In the edition of *The Cyanamid Magazine* published in the winter of 1963, American Cyanamid ran a series of magnified photos of insects under the title "Wanted: For Murder and Theft." The accompanying article claimed, "It has always been war between man and insects. Now, as both close in on a limited food supply, the battle intensifies. Total war. In the face of this, Rachel Carson suggests we disarm."[53] To further make its point, Cyanamid created captions that used the tone of a police blotter: "Jaws of Southern Armyworm are typical of the voracious maw of all the caterpillars who annually chew up a healthy percentage of the world's salads and vegetables." "A nuisance? More than that. The house fly tracks vermin all over your home and on your food. Larvae of other flies infect cattle. A cousin, the tsetse fly, has spread deadly sleeping sickness in Africa." And finally, "The Anopheles mosquito probes your skin with this wicked-looking instrument and feeds on your blood. It can carry the parasite that causes malaria. Another species transmits yellow fever this way." The threat came in many forms and was potentially everywhere.

Clearly, the magnified images and texts play off specific cues from the monster movies of American popular culture in the 1950s. One can see the "Creature from the Black Lagoon," "The Attack of the 50-Foot Woman," "Them," or the more famous nuclear experiment gone awry, "Godzilla," sharing center stage with the southern armyworm or the *Anopheles* mosquito. Defeating these technological nightmares, or "natural disasters," depending upon one's point of view, required the development of more powerful technologies, even though the misuse of manmade technologies had created the menace in the first place. Godzilla's demise, for example, came only after a reclusive scientist developed the "oxygen destroyer," an invention powerful enough to purge all the oxygen from the oceans.[54] One can only imagine the ecological fallout from that experiment. Similarly, the logic of postwar spray campaigns was to stay one step ahead of the evolutionary process, using the most modern chemicals, regardless of the ecological damage that might ensue.

It is highly doubtful that *The Cyanamid Magazine* purposely captured the paradox of chemical pesticides. Nor could one imagine that the magazine recognized that the solution to the pest problem might be as dangerous, if not more so, than the original problem. Yet what is important to consider here is that Cyanamid articulated a vision of the insect problem that lay outside U.S. borders. The "world's salads and vegetables" were at risk. Sleeping sickness endangered the lives of people on the African continent, not Americans. And malaria and yellow fever, while still a potential threat to Americans in 1960, had been virtually eradicated within the United States and were largely understood to be "tropical diseases." In scripting its attack on Rachel Carson, Cyanamid responded to the political fears of invasion from some unknown force looming beyond the U.S. border. In making this appeal, Cyanamid reinforced the boundaries between the United States and the international other, between the free and the unfree, between triumph over insects and surrender.

Not to be outdone, soon after sections of *Silent Spring* appeared in the *New Yorker*, Velsicol Chemical Corporation's secretary and general counsel, Louis A. McLean, sent a letter to William Spaulding, president of Houghton-Mifflin, publisher of *Silent Spring*, wishing "to call several matters to [his] attention from a legal and ethical standpoint." McLean's letter focused on what Velsicol claimed were the mistruths published in the *New Yorker* about the company's products chlordane and heptachlor. McLean threatened a lawsuit if these supposed misstatements were not corrected in or eliminated from *Silent Spring*.[55]

In the course of McLean's five-page letter, he alluded to a larger, potentially more dangerous threat that loomed beyond the nation's borders. He wrote, "Unfortunately, in addition to the sincere expressions of opinions by natural food faddists, Audubon groups and others, members of the chemical industry in this country and in western Europe must deal with sinister influences, whose attacks on the chemical industry have a dual purpose: (1) to create the false impression that all business is grasping and immoral, and (2) to reduce the use of agricultural chemicals in this country and in the countries of western Europe, so that our supply of food will be reduced to east-curtain parity. Many innocent groups are financed and led into attacks on the chemical industry by these sinister parties."[56] Who these "sinister" forces were remained unclear, although the construction of subversives out to destroy American capitalism and food production clearly suggests the ever-present communist menace. In defending itself from the controversy over *Silent Spring*, the chemical industry saw itself as a weapon against communism, protecting the home front and the American standard of living while ensuring an abundant food supply to overshadow that of the "east curtain."

At the annual meeting of the National Agricultural Chemical Association (NACA), held in New Jersey in September 1962, the major topic of interest, "to nobody's great surprise," *Chemical Week* reported, "was what to do about Rachel Carson."[57] The industry had already compiled a rebuttal to Carson's *New Yorker* articles, a thirteen-page pamphlet titled *Fact and Fancy*, which NACA members distributed to any interested party.[58] Emerging from the conference, however, NACA officials decided to meet Carson's challenge in two specific ways. First, it decided to take a "positive approach" to the public relations problem: "Rather than questioning Miss Carson's credentials the pesticide makers appear ready to defend themselves by pointing to their achievements."[59] By "accentuating the positive," the NACA leadership hoped to draw on international public health and agricultural development projects to demonstrate the beneficence of chemical pesticides. Second, the NACA recognized the need to inform American consumers about the bounty of American agricultural production. The industry's leading spokesperson, Dr. Robert White-Stevens of American Cyanamid, stated, "We, in agriculture . . . have researched ourselves into obscurity both politically and socially. . . . We have been so preoccupied in agricultural research, rural education and extension that we have essentially ignored the urban peoples in our communications, and have been talking largely to ourselves. We need to tell the urban peoples in a thousand places and a thousand ways what scientific agriculture, including agricultural chemistry, has meant to their health, welfare and standard of living."[60]

Not one to criticize the research methodology of his counterparts, White-Stevens nevertheless saw the limitations of industrial scientists. He recognized that laboratory research and fieldwork isolated scientists from the larger context of a particular problem, whether that problem was crop production or insect control. He also realized the changing political dimensions of the argument. For better or for worse, White-Stevens understood that the debate over DDT and other chemical pesticides was not simply about scientific expertise, but was about the validity of scientific knowledge within the public and political arena. In this public arena, the divisions over the risks and benefits of pesticides widened.[61]

The changing nature of the DDT debate undermined the claim that industrial scientists were holders of universal truth. No longer could industrial scientists proclaim the safety of a chemical pesticide, or any other chemical compound, for that matter, without being questioned by a concerned public. The authority of the expert had vanished and was replaced by a much more open and politically charged environment. So rather than grapple with the complexities of Carson's ecological philosophy, the chemical industry launched its own public relations campaign to promote the logic of industrial science to urban consumers and to those who supported the ambitions of American foreign policy to eradicate disease and eliminate famine.

For White-Stevens and other agricultural chemists, the "positive" referred to pesticides' maintaining the American standard of living, a lifestyle in which fresh and "attractive" food was always available to consumers. It also referred to pesticides' protecting Americans from foreign invaders and working to ensure that famine did not threaten the nation's interests abroad.

The issue of The Cyanamid Magazine from fall 1963 addressed both of these matters. First, the cover photograph depicted a worm-infested apple "speaking" to a group of beautifully ripe, highly polished apples. The small apple says, "Organic farmers love me. I was grown without pesticides, chemical fertilizers, or modern farming techniques." In unison, the other apples retort, "Obviously!" It seems beauty does come at a cost. Second, the corresponding article asks, "Will We Always Have Enough?" Cyanamid reported, "Americans are undeniably the best-fed people in the world; they are the best-fed nation ever to exist. . . . The food [Americans eat] has the highest nutritional value ever achieved. It is available the year around. It is full of variety. It is offered at the lowest relative price anywhere in the world."[62]

Still, the potential for this bounty to come to an end reflected two key issues: world population growth and the dissent of Rachel Carson. The world's population increased by an average of 20 percent per decade, rising from

2.5 billion to over 3.6 billion people between 1950 and 1970. Population experts warned that the rapid growth of human populations would have a catastrophic impact on the environment, particularly in regard to food production.[63] Under these circumstances, Cyanamid expounded, "No wonder food scientists feel the weight of responsibility. No wonder researchers race against time to perfect new fertilizers, weed killers, and insecticides." Carson, they argued, would simply reject the technical advances of modern industrial science and instead return to "the time when a man had no need for insecticides, when he went out alone to plant nine bean rows in a bee-loud glade." Obviously, for Cyanamid, the choice between past and future was clear: "We simply *cannot* have things both ways."[64]

Population growth, international food production, and the omnipresent communist menace provided a framework in which to promote the industry in the public eye. In this regard, the internationalization of the chemical industry was not only in the industry's self-interest, but in the nation's interest as well. Food and geopolitics were never far apart.

Following the tragic death of President Kennedy, Lyndon Johnson continued to advance humanitarian aid as a critical element of U.S. foreign policy. In 1964, Johnson delivered a message to Congress outlining his commitment to foreign aid. We would "be laying up a harvest of woe for us and our children," he claimed, "if we shrink from the task of grappling in the world community with poverty and ignorance." U.S. foreign aid was not simply a humanitarian endeavor, Johnson told Congress. Poverty and ignorance were "the grim recruiting sergeants of Communism."[65] For Johnson and his advisors, humanitarian aid and fighting the Cold War were linked. Nearly two-thirds of USAID aid from 1963 to 1968 went to India, Pakistan, and Turkey, nations that were "judged to be of the greatest concern to U.S. national interests."[66]

In 1966, after a series of droughts that reduced crop yields, yields that had been the model of Green Revolution success, India faced catastrophic shortages of basic grains, requiring it to import nearly 10 million tons of grain. Members of the Johnson administration expressed enormous concern. Secretary of Agriculture Orville Freeman claimed that the Indian government was in "deep trouble" with the food issue.[67] Walt Rostow stated, "No greater threat exists to the long-term interests of the United States than the growing food crisis in the less developed world."[68]

The massive crop losses came at a time when population experts were sounding the alarm over the dramatic rise in human population. The looming food crisis, they believed, could dramatically alter the political situation in developing nations, thus complicating an already dangerous and divided world.

To respond to this political threat, President Johnson launched an ambitious aid project in 1966 called Food for Freedom. The program was designed to expand U.S. commitments to international aid for Green Revolution technologies and to provide direct food assistance to countries unable to provide their citizens with basic grains. Yet these short-term solutions were merely stopgap measures. The program's long-term goals were to ensure that countries receiving aid embarked on a program of "self-help." Upon signing the program into law, Johnson claimed, the "program will provide American food and fiber to stimulate greater productivity in the developing countries . . . [but] self-help measures are being taken by the recipient [countries] to improve their own capacity to provide food for their people."[69] Another Johnson aide, Donald Horning, put the matter more directly: "The problem of feeding the world is plainly a human use of science and technology."[70]

In a span of a few years, the deployment of the "technological package" radically transformed rural areas worldwide. In Asia, for example, areas planted with Green Revolution technologies increased from 200 acres to 38 million acres from 1965 to 1968.[71] In Latin America, Green Revolution technologies continued to spread throughout the region, as Mexican seeds flowed south to Colombia, Chile, and Argentina. And, as always, chemical fertilizers and pesticides followed the global flow of rice and wheat seeds.

Indeed, as one study indicated, "in the 1950s, [Indian] farmers had little incentive to apply fertilizers." But by the mid-1970s, fertilizer use increased sevenfold for rice production alone. USAID promoted the private-sector distribution of fertilizers and pesticides. According to one report, "In Pakistan, the use of fertilizer has risen from 30,000 nutrient tons in fiscal year 1960 to 270,000 nutrient tons in fiscal year 1968."[72] In 1952, India began producing benzene hexachloride, an organochloride similar to DDT, in a newly constructed facility in Calcutta, increasing its production capacity as a way to offset the cost of importation. USAID also helped the U.S.-based conglomerate Bechtel to begin negotiations with India to construct five fertilizer plants. Although the project ultimately failed, the ways USAID actively promoted private funds to assist a public endeavor reflected the State Department's commitment to the market economy as a central element of the development project. Other parts of Southeast Asia sought to encourage foreign investment as well. In Indonesia, for example, facing a shortage of foreign exchange, the government invited foreign manufacturers to supply farmers directly with chemical fertilizers and pesticides.[73]

For U.S. policymakers, the belief in technological solutions to complex social and economic problems determined political and economic strategies for reshaping the world. In theory, technological solutions provided much-needed relief to those affected by malnutrition and disease, while providing the mechanisms to develop "self-help" programs. Yet, in applying these solutions, development experts cast aside different forms of scientific knowledge, particularly those rooted in local communities, in order to dispense the latest technological innovation. The social, cultural, economic, and political impacts of this agricultural model were profound, and they continue to shape the politics of the global food system today.

Technology afforded a developing nation a means to project an image of itself as modern to the world community. The "Mexican miracle" started under a cooperative agreement between the Mexican government and the Rockefeller Foundation, but it took on greater significance as the technical package to improve crop yields underwrote the capitalist expansion of the country. With the 1968 Olympic games—a sure sign of late twentieth-century modernity if there ever was one—Mexico became a symbol of success for the development project. In India, one only has to look at advertisements for "Modern" bread to fully appreciate the ramifications of the development project. Produced by Modern Bakeries Limited, an Indian firm established in 1965, the bread is "vitaminised" and chemically fortified with Lysine, an essential amino acid usually found in lentils, chickpeas, and soybeans. The loaf also looks remarkably like Wonder Bread, a product of a company that during the 1940s became one of the first manufacturers to fortify bread with vitamins and minerals. While fairly insignificant in a country as populous as India, Modern bread nevertheless represented a significant shift in the ways in which the country thought about and marketed food. As historians Frederick Cooper and Randall Packard suggest, development was "fundamentally about changing how people conduct their lives."[74] Indeed, the technology transfer of the development project radically transformed the bodies, lives, and economies of many developing nations. While much could be said about the benefits, they also came at a high price.

Not only had chemical corporations come to the defense of DDT in the wake of *Silent Spring*, but states increasingly looked to private industry to support their modernist agendas. Throughout the decades of development, U.S. corporations had become one of the primary drivers of this technology transfer. Freedom from pestilence, it seemed, meant steadily embracing free-market principles.

Within the United States, however, the development model undermined

Advertisement for Modern bread, Bombay (Mumbai), India.

the ecological claims made by Rachel Carson. For Americans, the deployment of chemical pesticides overseas seemed like a minor nuisance compared to the potential threat of communist advances. Indeed, the very science that Carson claimed caused such destructive harm in the United States, when applied elsewhere, seemed to protect American lives. Of course, absent from this assessment was a consideration for those whose lives were transformed by the Green Revolution.[75]

Yet the promise of technology called into question how best to apply it in a global context. Surprisingly, the answers were not entirely clear. Even for advocates of technical aid, the complexity of the food crisis and population growth proved quite vexing. In 1966, agriculture secretary Orville Freeman and Lester Brown, a USDA official who later founded the World Watch Institute and the Earth Policy Institute, wrote a report on the use of agricultural technologies in the struggle to reduce famine. A draft of the report imagined an alien landing on earth to survey the application of science and technology in the modern world. The authors surmised that the alien in question would proclaim, "Modern science and technology were designed by some evil genius to reward the affluent and punish the hungry." The reason for this assessment lay at the paradoxical heart of the technological dilemma. "For the West," Freeman and Brown wrote, "the past 67 years have seen the culmination of an agricultural revolution which has, for the first time, made it possible for whole nations to have assured food supplies. But in Asia, Africa and Latin America, modern science's greatest impact has been to reduce death rates. The result has been a population explosion without historical precedent, an expansion in human numbers demanding more and more food just to avert mass starvation."[76] Indeed, the massive spray campaigns that reduced malaria rates to historic lows were creating another equally difficult problem, increased famine. These issues would prove troublesome for development experts, and for the emerging Green Revolution in the United States as well.

In 1966, Jamie L. Whitten, a powerful Democratic congressman from Mississippi and chair of the House Appropriations Subcommittee on Agriculture, released his only book, *That We May Live*. The book evolved out of Whitten's request to the surveys and investigations staff of the Appropriations Committee to "conduct an inquiry into the effects, uses, control, and research of agricultural pesticides, as well as an inquiry into the accuracy of the more publicized books and articles which increase public concern over the effects of agricultural pesticides on public health."[77] More specifically, he was interested in the impact of *Silent Spring* and DDT. Whitten's interest in DDT grew out of the needs of his own constituency, the cotton industry. Southern cotton farmers were some of the biggest users and supporters of chemical pesticides. By the late 1960s, nearly 70 percent of the domestic consumption of DDT occurred in cotton production. First elected to Congress in 1941, Whitten amassed a considerable amount of power and exercised enormous influence over most of the nation's farm legislation in the postwar period. He used his political muscle to ensure that any pesticide regulation would not interfere with his constituency's livelihood.[78]

The findings of the congressional hearing were published as *The Whitten Report*. Not surprisingly, the report offered a powerful critique of *Silent Spring*, questioning, once again, Carson's scientific credentials and her capacity to understand the enormity of the agricultural dilemma. Because it refuted Carson's findings, *The Whitten Report* was subsequently reprinted by the NACA as *That We May Live*. It was purely industry propaganda, even though it emerged from the halls of Congress. Once again, however, Whitten connected Carson to national defense. He wrote, "In our fight with Communism our agriculture is perhaps our greatest asset. Its scope embraces both national security and national health."[79]

Wheeler McMillen, a freelance writer and former editor of the *Farm Journal*, made perhaps the most direct link between pesticides and the spread of communism overseas. In his 1965 book, *Bugs or People?*, McMillen cast the question in simple terms. (During the war on communism, nuanced arguments were taken as a sign of communist sympathies on behalf of those making them.) McMillen framed the pesticide issue as purely an all-or-nothing situation. In a chapter called "The World's Hungry People," McMillen compared the life of a sick, malnourished child from some undetermined Third World country to that of a strong American child who benefitted from pesticides and industrial agriculture. McMillen wrote:

He may not have a future, for throughout his growing years he will be threatened by diseases, many of them insect-borne, that are prevalent in his area. If he does achieve adulthood he will not likely ever be able to exert the physical energy that is required by what in this country we think of as an honest day's work.

He will resent his condition. He will wonder why fate has dealt with him so badly. When the demagogues come, as they will, he will hear their words. Being ignorant, he may believe. Millions like him will believe—believe the wrong things, things that are not true. They will learn to hate those who wish to and can help the most.[80]

Fear that millions of people could learn the wrong things and fall under the demagoguery of communism, all because access to American pesticides and agricultural technologies were denied, elicited McMillen's anticommunist rant. In his chapter "Guerrillas of the Underground," McMillen explored the subterranean world where grubs, worms, and beetles feast on the very foundation of agricultural production, the root system. Corn-seed beetles, corn rootworms, and click beetles, McMillen contended, "spend their hidden lives damaging crops, reducing yields, gnawing away farmers' profits, and subtracting from human food supply."[81] The subversive nature of these "underground pests," McMillen argued, made the potential for crop failure much more likely, since, like communists, bugs were out to destroy the very fabric of the nation.

That We May Live and *Bugs or People?* created a discursive response that manufactured a foreign "other" that was helpless to stamp out pests and communists (in fact, McMillen's publishers reprinted Monsanto's "Desolate Year" in the appendix of *Bugs or People?*). As one plant scientist rhetorically stated, the chemical industry believed "that salvation lies at the end of a spray nozzle."[82]

Hunger, however, was not just a Third World problem. In 1968, CBS broadcast a special report called "Hunger in America," which exposed the inadequacies of the industrial food system and the failure of federal food aid. The program had a profound impact on the nation, leading many citizens to question the allocation of federal dollars for international aid. Already politically wounded from the deterioration of the Vietnam War, President Johnson received the brunt of the criticism. "How do we as a country justify spending three billion dollars a month in Vietnam," one angry citizen wrote, "and permit ten million fellow Americans to suffer the ravages of malnutrition." Another letter remarked, "Eliminate Hunger in America first." And another concerned citizen simply stated, "Dear Mr. President: I have just watched CBS Program Hunger in America and cannot understand why our government does not eliminate foreign aid and take care of America first."[83]

As Americans grew increasing frustrated with the quagmire in Vietnam and the federal government's inability to effectively deal with domestic issues such as hunger and civil rights, foreign aid became a source of contention. At the beginning of the 1960s, the logic of development seemed clear and resolute. By the end of the decade, international development seemed costly and capricious. Throughout the decade, however, the chemical industry continued to promote the benefits of industrial agriculture and the Green Revolution to counteract the negative publicity caused by the publication of *Silent Spring*.

IN DEVELOPING A "POSITIVE" RESPONSE to *Silent Spring*, the chemical industry, with the assistance of like-minded critics, did more than simply extol the benefits of industry chemicals. They also tied their rhetoric to the struggles of the Cold War, a war fought on two fronts. The domestic fight sought to defend the agricultural and public health interests of the nation from unwanted invaders. The "home front" served as a powerful metaphor that encouraged Americans to take up the spray gun against insect pests. Yet the "offensive" assault on insects overseas proved to be just as vital, if not more so, since U.S. strategic interests were beyond U.S. boundaries, away from domestic fallout shelters. The construction of the insect (and communist) menace relied on the ominous threat existing "out there." By accentuating the positive aspects of using pesticides overseas, the chemical industry allied itself with the foreign policy strategies of international development, containment, and the spreading of democratic capitalism to the underdeveloped world. In essence, pesticides, and the chemical industry by extension, were engaged in the battle against communism, protecting national interests while expanding democratic principles.

Carson, on the other hand, effectively demonstrated the pervasiveness of

DDT. DDT was, she contended, a borderless poison that had serious consequences far and wide. Yet for Carson, the poisoned landscape she so aptly described was iconically American. She wrote about the "sagebrush lands of the west" and "Hawk Mountain in eastern Pennsylvania." She introduced readers to California's "Clear Lake," a place that "has long been popular with anglers." And she wrote about the small suburban communities and New England towns that were subjected to unwanted aerial spraying.[84] These quintessentially American locations were sites where Carson exposed the real danger of chemical pesticides. They were pristine and endangered. As a nature writer, Carson placed herself squarely within a long tradition of writers who romanticized the American landscape and the "balance of nature."[85] So, despite Carson's forceful argument that DDT was pervasive, persistent, and global, her portrait of a nation under assault encouraged a domestic Green Revolution that sought to protect national lands, without much consideration for the world beyond the border.[86]

The publication of *Silent Spring* and the scientific response to it were only the beginning of the debate over pesticides. As the debate shifted from the public realm of books, magazines, television shows, and industry journals into the bureaucratic maze of federal regulatory agencies, the meaning of DDT would continue to be shaped by its domestic and foreign use.

6

IT'S ALL OR NOTHING
Debating DDT and Development under the Law

*Died. DDT, age 95, a persistent pesticide and onetime humanitarian. Considered
to be one of World War II's greatest heroes, DDT saw its reputation fade after it
was charged with murder by author Rachel Carson. Death came on June 27 in
Michigan after a lingering illness. Survived by dieldrin, aldrin, endrin,
chlordane, heptachlor, lindane, and toxaphene. Please omit flowers.*
—Hal Higdon, "Obituary for DDT (in Michigan)," 1969

*If I must make a choice between some unimportant
species of bird and man, I'll vote for man.*
—Norman Borlaug, 1971

Over the course of the contentious debate touched off by *Silent Spring*, the
WHO's global malaria eradication program appeared to be at a crossroads.
Despite the stunning reversal of malaria rates in large parts of the world,
skyrocketing administrative costs, funding cuts, mosquito resistance, and,
perhaps more importantly, the realization that eradication of the disease was
more fantasy than reality placed the global campaign in a precarious position.
"We have reached a point of no return," malaria specialist Leonard Bruce-
Chwatt commented in 1969, "and we must persist in our efforts to eliminate
the major endemic diseases, even if our strategy must be changed." Indeed,
because of structural and ecological problems, many health experts and government officials began rethinking eradication. Later that year, the WHO
called for a suspension of the global eradication program, bringing the once
revolutionary project to an ignominious conclusion. It was clear to most that
eradication had failed, resulting in economic, political, and environmental
costs that were staggering. The WHO "had used its medical professionalism
and its position as the international health authority," historian Amy Staples

concluded, "to foist upon the most needy countries of the world a plan flawed in both design and execution."[1]

The development decade quickly turned into a "decade of disappointment." In the context of mounting concerns about population growth, the optimism of the previous decade about the power of technology and the promise of development faded. In 1972, the *Washington Post* exposed the hidden costs of the Green Revolution, a revolution that "accelerates the flight from the land, . . . exacerbates differences between farmers who profit from it and farmers who don't . . . [and] puts more food on the market but does not automatically provide the poor with the resources to acquire that food." In light of these developments, the former USDA administrator and onetime advocate of the Green Revolution, Lester Brown, commented that the technological package "of better seeds and cultivation techniques is not in itself a solution to the global problem." Indeed, the promise of the Green Revolution, a pillar of postwar development schemes, "turn[ed] brown."[2]

Amid these international problems, the debate over the meaning, regulation, and use of DDT within the United States grew more contentious. Not only had *Silent Spring* raised concerns over the pesticide, leading to highly charged disputes within the scientific community, but, encouraged by Carson's ecological ideas, environmental organizations began to politicize the pesticide issue, much to the dismay of the chemical industry. Foremost among these groups was the Environmental Defense Fund (EDF), an organization founded in 1967 that relied on ecological science and the law to change public policy.[3] As one of the EDF's founding members, Dennis Puleston, once observed, "There are countless groups which have been relying too much on public outcry, petitions and newsletters."[4]

In bringing this new form of politics into the fold, the EDF challenged ideas about U.S. environmental regulation and law and fashioned a more adversarial approach to the DDT question. Not only would this approach quicken the regulatory process within the United States, but it would sharpen the lines between technological enthusiasts and proponents of ecology. So while the EDF initiated what was primarily a domestic judicial process, the impact of that process had far-reaching implications that redefined scientific knowledge within and beyond the United States. To the dismay of many Americans, the contest over DDT also shed light on the changing relationship the nation had with the rest of the world. The marvel of a modern technology that could rid the world of pestilence and disease had been replaced by a chemical pesticide that could severely damage the world's ecosystem. Though DDT was one of the critical technologies of American postwar expansionism, the chemical exposed a fundamental flaw in the logic of the American Century, the limits of technology's ability to solve complex social, cultural, economic, and political problems. And the contentious struggle over DDT was more than a domestic regulatory issue; it was also a contest about science, technology, and American influence abroad.

IN 1967, THE EDF ASKED a federal court to stop fifty-six Michigan cities and towns from using DDT to combat Dutch elm disease.[5] Another EDF suit filed in state court challenged the Michigan Agriculture Department's spraying dieldrin on fruit trees threatened by Japanese beetles. Led by an outspoken and brash attorney, Victor Yannacone, and a team of scientists—George Woodwell, Dennis Puleston, and Charles Wurster—the EDF filed lawsuits in Michigan, an attempt to open a public forum on DDT, which, the EDF's founders believed, would end the controversy once and for all. While both cases were dismissed on jurisdictional grounds, fifty-five Michigan towns voluntarily cancelled their planned sprayings. The EDF's legal actions single-handedly changed the dynamics of pesticide politics in the United States.[6]

The legal strategy also set the EDF apart from other environmental groups that came into existence or expanded during the radicalism of the 1960s. Greenpeace, Friends of the Earth, and even the staid Sierra Club, which temporarily lost its tax-exempt status because of its "radical" politics, engaged in direct political action and sought the expansion of a politically astute constituency as the basis for change.[7] In contrast, members of the EDF were committed to working within the system, ensuring that existing laws were being upheld as they worked to create new environmental laws. Amyas Ames, who served as an EDF trustee during the early 1970s, commented, "It is the policy of E.D.F. to work within the legal-executive-economic frame of our society rather than to obstruct or impede action necessary for the community; to ask that protective laws be observed, that environmental standards be set before, not after a project is undertaken and that projects be designed to minimize harmful impact on the environment."[8]

At a time when street riots occurred, in the long, hot summer of 1968, Vietnam protests spilled out onto the streets of Washington. Racial, gender, and class politics became increasingly radicalized, and the EDF's approach to environmental politics seemed rather mundane, even though others saw those politics as a sign of "a growing militancy in the conservation movement."[9] The civil rights movement, which used the courts to seek redress for political, social, and economic segregation, was not unlike the EDF's movement, which embarked on a similar mission to expand federal protections under the law.[10] Not surprisingly, Yannacone, the EDF's counsel, had tried cases on behalf of the NAACP and understood the power of the courts and the potential for creating a new body of environmental law.

Sure enough, these legal maneuvers did not go unnoticed. Just ten weeks after the EDF had formed, the editors of *Farm Chemicals* sounded the alarm about the new environmental group, declaring, "Storm clouds are gathering over the pesticide industry, and the group seeding them is the Environmental Defense Fund."[11] Given the industry's response to *Silent Spring*, it came as no surprise that Gordon L. Berg, *Farm Chemicals* editor, condemned the environmental group as a "nature cult," "food faddists," an "organic farming cult," and a "pesticide-wildlife cult." Fully cognizant that the EDF had set its sights on DDT and "the elimination of chlorinated hydrocarbons," *Farm Chemicals* told its readers, "After that is accomplished, another group of pesticides will come under systematic attack."[12]

Parke C. Brinkley, then-president of the National Agricultural Chemicals Association (NACA), reaffirmed this belief. "It wouldn't make any difference if they managed to get the chlorinated hydrocarbons banned," Brinkley asserted; "it wouldn't do a thing in the world except charge them up and [the] next step would be the carbamates and the organophosphates."[13] The EDF did little to assuage these fears. Yannacone remarked that once the battle against DDT was won, other pesticides would "fall like dominoes," a particularly powerful metaphor at the height of the war in Vietnam.[14]

Despite Yannacone's vociferousness, Berg was also acutely aware, as many in the "hard" sciences were, of the growing popularity of ecology within mainstream America, a situation that both frightened and angered many. Berg informed his readers, "The character of the scientific community is changing." Rather than being a positive change, Berg proclaimed, it was a "sign of the times" in an "age of opportunism," suggesting the radicalism of the 1960s had somehow infiltrated the hallowed halls of industrial science. "We cannot afford to listen to these opportunists," Berg fumed, "while more than 12,000 die each day of starvation and malnutrition and millions of others cry out for our help."[15]

Undeterred, the EDF forged ahead in its battle against DDT, moving from its muted success in Michigan to a new battleground, Wisconsin. Citizens had the right under Wisconsin law to petition state agencies and ask for a "ruling of applicability" on the facts a particular agency used to enforce the law. At the judgment of the agency, a public hearing would be held, which would be "subject to the rules of evidence and cross-examination." In this case, the EDF, on behalf of the Citizens National Resources Association of Wisconsin and the

Wisconsin Izaak Walton League, petitioned the Wisconsin Department of Natural Resources to determine whether DDT was a pollutant under Wisconsin statutes. The petition, which focused on a small part of Wisconsin's water pollution laws, Wisconsin statute 144.01, questioned whether or not DDT was a pollutant and, if it were, whether its use would be in violation of the state's water pollution laws.[16] "This seems to be the time," the biologist and EDF founding member Charles Wurster avowed, "to absolutely bury DDT scientifically, to settle this issue once and for all and eliminate any further talk about this just coming from an emotional bunch of bird watchers."[17]

The hearing began on December 2, 1968, and would last for five months (although it was interrupted by two lengthy recesses), calling thirty-two witnesses and creating over 2,800 pages of testimony. While the findings of the judicial proceeding would be quite narrow, the scope of the testimony revealed a much broader set of issues that were, at times, regional, national, and global. Due to the rules of cross-examination, the chemical industry and environmentalists were forced to confront the relative strengths of each other's arguments. For example, for the first time in a rule-making proceeding, the EDF and environmental scientists could directly challenge the safety claims made by the chemical industry. Conversely, armed with powerful statistics about DDT's life-saving capabilities, the chemical industry directed environmentalists to address questions about the successes of malaria eradication and control projects overseas.

In a larger sense, science was on trial. The president of Montrose Chemical Corporation (the largest U.S. manufacturer of DDT), Samuel Rotrosen, stated, "I don't say these people aren't true scientists, but their interest appears to be 'let's worry about birds more than people.' I think we'll have qualified scientists who will put the picture in perspective."[18] Rather than being a simple rule-making proceeding at the state level, the Wisconsin hearing underscored conflicting notions of scientific knowledge that defined and would continue to shape the meaning of the American Century.

To support its case, the EDF called on ecologists, entomologists, botanists, ornithologists, and zoologists. The wide-ranging disciplinary spectrum was a key component of the EDF's case. Since its claim rested on the belief that DDT's persistence and mobility produced massive ecological problems that could not be determined by narrow scientific evidence, the EDF attached great importance to multidisciplinary thinking and an ecologically minded approach. "A basic element of the EDF position," Wurster remarked, "has been to emphasize the complexity of the environmental sciences and the need for interdisciplinary cooperation, then to demonstrate through cross-examination that the DDT proponents are narrow specialists out of touch with the rest of the scientific community."[19]

To highlight the political and environmental importance of the hearing, Yannacone called Senator Gaylord Nelson as his first witness. First elected to the Senate in 1962, Nelson was a committed environmental leader, having introduced or sponsored a number of key pieces of legislation prior to his testimony, including legislation preserving the Appalachian Trail and the Wilderness Act of 1964. By 1970, Nelson would be better known as the originator of Earth Day, an environmental teach-in that was transformative in the modern environmental movement.

At the hearing, Nelson's testimony affirmed the ecological extent of the pesticide problem. "Through [the] massive, often unregulated use of highly toxic pesticides, such as DDT and dieldrin," Nelson maintained, "the environment has been polluted on a worldwide basis. In only one generation, these persistent pesticides have contaminated the atmosphere, the sea, the lakes, and the streams and infiltrated the tissues of most of the world's creatures, from reindeer in Alaska to penguins in the Antarctic, including man himself." And as if to portend the debates to come at the federal level, Nelson claimed, "I have introduced legislation in the Senate to prohibit the interstate sale or use of DDT in the United States. Just as this hearing is solely concerned with DDT, my bill would not affect the sale or use of any other pesticide. It should be made perfectly clear that we are not seeking a ban on all pesticides, only on the most persistent, most dangerous one—DDT."[20] The question on the minds of many observers was whether DDT was, in fact, a dangerous pesticide.

Building on the EDF's ecological case, Hugh H. Iltis, a botanist at the University of Wisconsin, Madison, testified to the ecological interdependence of Wisconsin's environment. Iltis introduced the idea of "regional ecosystems" as a way to draw attention to the interconnectedness of the natural world. His work showed "that Wisconsin is not an island; and whatever occurs elsewhere in the world, in particular with DDT, occurs to some extent in Wisconsin; and that whatever occurs in Wisconsin, has some effects outside."

The molecular biologist Robert Risebrough from the Institute of Marine Resources, University of California, Berkeley, provided evidence that DDT was, in fact, a worldwide problem. His research on marine food chains in the Pacific and Indian Oceans highlighted the itinerant nature of the pesticide. Under questioning, Risebrough claimed that DDT and its metabolites were "the most abundant synthetic pollutant[s] in the global system."[21]

On the second day of the hearing, Yannacone called Wurster to the stand to testify on the environmental consequences of DDT. Drawing on a large body of scientific evidence, Wurster presented a well-formulated attack, claiming the environmental problems stemmed from DDT's broad biological activity, its persistence, mobility, and solubility. In his capacity as an expert witness and an EDF scientific adviser, Wurster forcefully argued that DDT should not be used under any circumstances as an agricultural pesticide or in response to any public health threat, domestically or abroad, a position that would marginally change over time.

Although Wurster's direct examination lasted one day, his cross-examination by Louis A. McLean, a former Velsicol Chemical Corporation attorney and a dedicated defender of DDT, would last three days. Taking on a range of matters, from the details of particular research studies to Wurster's political work on behalf of the EDF, Wurster and McLean engaged in a lively debate over the risks and benefits of DDT. McLean asked, "Would you make a value judgment in connection with the application of pesticides to protect public health?" Wurster responded, "On the one hand we are faced with the situation where DDT is doing great worldwide damage. On the other hand we have the opportunity to use many alternatives for public health insect problems. It is my suggestion that we make use of those alternatives which are not so dangerous as I said. There is no further scientific justification for the continued use of DDT under any circumstances where it can be released in the biosphere."[22]

McLean asked whether Wurster was aware that DDT was still used for malaria control in "tropical countries." Wurster responded, "It would be best if it were not used," later adding that DDT was "used in very large quantities." McLean continued his questioning, directing Wurster to respond to the WHO's practice of indoor house-spraying. Wurster suggested that the programs ignored the environmental hazards of DDT application. "In fact," he stated, "it's just used willy-nilly, in many cases without really demonstrating its mode of action."[23] Asked whether the cost of DDT was one of the essential characteristics of its overseas successes, Wurster countered, "The price we are paying in environmental damage for DDT is so high that we can't afford to pay that price. Yes, I'm consider[ing] price. That's one of the reasons we can't use DDT anymore."[24]

McLean's questioning had particular pertinence because of shrinking WHO and U.S. foreign aid budgets. In 1968, for example, DDT sold for around 17 cents a pound, while malathion, a possible substitute, sold for anywhere between 68 to 95 cents a pound.[25] Uses of alternative pesticides for malaria control, moreover, required frequent application, thus raising the cost even higher. McLean, however, neglected to mention the failure of the WHO malaria eradication program or the evolutionary process that produced pesticide resistance.

Scientists on both sides of the issue provided evidence in an attempt to sway the judicial outcome and public opinion. Ecologists warned against the long-term consequences of DDT, while industrial scientists emphasized the significance of the pesticide for international agriculture development and public health projects. The hearing also reflected the growing complexity of environmental policymaking, as scientists argued over the hazards of chemical residues and exposures in the parts per million. While the chemical age had brought with it scientific complexity, this complexity was also difficult to regulate and, for the average American, nearly impossible to comprehend.

Having scientists debate the health and environmental consequences of substances in such small quantities was confusing enough. But having government agencies responsible for regulating the myriad compounds was nearly an impossible task. Indeed, because of the scientific nature of the evidence, much of this technical testimony was lost on the American public. But, perhaps more importantly, it became apparent that the role of the "expert" was changing. No longer were technical scientists holders of a universal truth. Rather, the Wisconsin DDT hearing highlighted an important shift in the role of science during the American Century. Not only could ecologists understand the scientific evidence presented by the chemical industry, but they could put it in a larger context that many people could comprehend. The judicial findings would ultimately support the EDF's position, although they would not be published for another year. Nevertheless, the EDF left Wisconsin confident of its ability to reframe the scientific arguments about the many uses of DDT.

The debate between ecologists and industry scientists was not settled in Wisconsin, however. Nor would it be settled in the years to come. But as the legal campaign against DDT moved from the state to the national level, the significance of the scientific debate would become clearer and increasingly politicized.

In 1969, the EDF initiated proceedings against the secretary of agriculture, Clifford M. Hardin, asking the USDA to suspend the registration of DDT, which would, in effect, immediately stop all DDT use.[26] Additionally, the EDF appealed to the Department of Health, Education, and Welfare (HEW), asking it to establish a "zero tolerance" policy for raw agricultural commodities. According to the EDF, "The petition requests FDA to lower the tolerance limits of DDT residues in human foods to zero, based on a law stating that carcinogenic materials cannot be tolerated in human food."[27] At issue were DDT's possible connections to cancer. Both petitions sought a remedy through the Delaney Amendment to the Federal Insecticide, Fungicide, and Rodenticide Act of 1947 (FIFRA), which was one of the few pieces of legislation in the postwar era that dealt with chemical additives in U.S. food production. Passed in 1958, the Food Additive Amendment—or Delaney Amendment—required the FDA to set a zero tolerance level for any chemical that caused cancer in man or animal. The EDF contended that DDT's persistence functioned as a chemical additive and its "carcinogenic" effects on mice warranted immediate suspension of the chemical's use on agricultural products. Reliance on the Delaney Amendment in both petitions, historian Thomas Dunlap has suggested, "marked a shift in EDF's tactics, as the reliance on the cancer issue indicated a change in the argument against DDT."[28]

Despite rising fears that DDT might be carcinogenic, the domestic regulation of chemical pesticides proved to be extremely lax. In September 1968, the General Accounting Office issued a report charging that the USDA failed to adequately manage pesticide regulations under FIFRA. Furthermore, the law requiring the secretaries of HEW and the Department of the Interior to produce evidence denoting which pesticide caused environmental or health hazards was being challenged, as was the close relationship between the USDA and the pesticide industry. An amendment to FIFRA in 1964 transferred the burden of proof of chemical toxicity from the USDA to the manufacturer, while making cancellation much easier for the USDA.[29] Congress began its own investigation into the USDA's handling of pesticide regulation in June 1969.

In April 1969, HEW secretary Robert Finch appointed a special commission to gather all available evidence on the benefits and risks of using pesticides. Chaired by Emil M. Mrak, the chancellor of the University of California, Davis, and a noted scholar on food science and technology, the Mrak Commission issued its report, which was "the most comprehensive study to date on pesticides and their impacts," in November.[30] Because of DDT's ecological impact, the Mrak Commission recommended a gradual suspension of the pesticide in the United States within two years, save for the uses "essential to the preservation of human health or welfare." Conversely, the commission held, "The risk-benefit considerations [for the use of DDT] differ somewhat from country to country depending on the particular problems encountered. Both benefits and hazards of using a pesticide must be evaluated carefully in order to determine the appropriateness of use in a given area."[31] Mrak's experience with food production problems in the developing world may have shaped his judgment about the effectiveness of the pesticide in some parts of the world, while he acknowledged the environmental damage at home. Because of its nuanced interpretation, the Mrak Commission report "served the purposes of neither side in the dispute."[32]

Accepting the commission's findings, Secretary Finch announced a two-year phasing out of all but the "essential uses" of DDT. Finch's concern over DDT grew during the summer, when the National Cancer Institute reported that a recent study "provide[d] substantial evidence to link DDT pesticides with carcinomas."[33] Faced with mounting public pressure as well—the National Audubon Society launched a campaign for a nationwide ban—Finch declared, "Prudent steps must be taken to minimize human exposure to chemicals that demonstrate undesirable responses in the laboratory at any level."[34] The announcement had far-reaching ramifications since Secretary Finch, USDA secretary Hardin, and Secretary of the Interior Walter Hickel earlier agreed to put into effect the recommendations of the Mrak Commission. In response, Hardin reluctantly announced a ban that would begin in thirty days on DDT's use on shade trees, on tobacco plants, in residential areas, and in aquatic areas. This was a modest concession at best, since more than 70 percent of all DDT used domestically was for cotton production. Hickel would later ban the use of DDT and other hydrocarbons on more than 500 million acres of federal land.[35] While federal agencies moved closer to a ban, or at least reducing the amount of DDT used in the United States, they challenged the standing of the EDF as a petitioner. The USDA, for example, claimed the EDF and other group petitioners lacked the legal standing to complain about the agency's actions since they were not directly or adversely affected by the agency's actions the way pesticide manufacturers and agricultural users were. Yet in a world where DDT was everywhere, the claim of direct effect had little substance. More importantly, the jurisdictional complaint lodged by the USDA reflected the narrowness of the federal regulatory apparatus regarding chemical pesticides.

Events outside the Washington beltway had a profound impact on the political struggle over DDT. In March 1968, the FDA confiscated 22,000 pounds of coho salmon from a Lake Michigan fishery because the salmon contained excessive levels of DDT and dieldrin, up to 19 parts per million (ppm), which was significantly over the 7 ppm limit set for other food products. In April of that year, Sweden announced it would impose a two-year ban on DDT, in order to discover if the ban would reduce the amount of DDT in plants and animals. Sweden also banned dieldrin and aldrin. Australia declared it was "rapidly phasing out" the use of DDT, to avoid possible restrictions of its produce in the export market.[36] In 1967, for example, U.S. officials refused to import 300,000 pounds of beef from Nicaragua because excessive DDT levels were found.[37] More significantly, the Australian regulation triggered a growing concern among the world's food producers for the international market during this period of tremendous environmental regulation. The Australian regulation called attention to the interconnectedness of pesticide policies on a national level and the flow of food commodities on the international market. Foreshadowing events and regulatory problems that would arise in the following decade, Australia's move to reduce the use of DDT makes it clear that U.S. pesticide regulation had considerable consequences for other nations.

After the Wisconsin DDT hearing, the EDF underwent a dramatic transformation as well. Victor Yannacone, the EDF's outspoken attorney, left the organization after a falling out with the other founding members over the future direction of the EDF. Membership dollars began flooding into the group's Long Island office (the number of contributors rose from 10,000 in October 1970 to over 26,000 a year later), along with a $285,000 grant from the Ford Foundation, which also funded Green Revolution programs around the world.[38] In response to criticism of the grant allocation, Ford Foundation president McGeorge Bundy stipulated, "In making a grant to EDF the Foundation did not prejudge that EDF's point of view was correct, but only that it was legitimate, important, and in need of external funding. If as a people we are successfully to protect our environment, we cannot afford to overlook any pertinent aspects of the impact of technology or muffle any responsible views."[39] The influx of funds went toward establishing regional offices in Washington, D.C., Denver, and San Francisco, hiring a new executive director and attorney, and developing a broader environmental agenda. This new professional entity reflected the maturation of a postwar environmental movement that grew from small citizen organizations concerned about local and regional problems to full-fledged professional lobbying organizations that saw Washington as the center of environmental policymaking on a national scale.

Alarmed by the influence of the EDF legal campaign, another group of citizens and scientists joined forces to shift the public debate. Formed shortly after the Wisconsin DDT hearing, Sponsors of Science—also known as DDT SOS—was "a voluntary association of professors, educators, scientists, and educated persons with experience in, or knowledge of, the research regarding proper and safe uses of pesticides, and especially DDT, in the areas of public health, medicine, and agriculture." More of an advocacy group than a committed team of researchers, the SOS worked to "correct the dangerous imbalance of publically expressed opinion regarding pesticides and DDT." From its perspective, the DDT debate had become one-sided and the "fairness doctrine," which required broadcasters and other news outlets to provide balanced coverage on controversial issues of public importance, "had been deliberately and completely ignored by most networks and local stations in the current controversy over pesticide use; and dangerously ignored in respect to the continuing need for DDT."[40] The SOS wanted to demonstrate the "proven human benefits from the use of DDT," which "far outweigh the unproven speculations of harm." Indeed, the SOS viewed DDT through an anthropocentric lens, as most development experts and chemical industry representatives did: "Human lives," "human health," and "our food supply" framed the narrative that made DDT use "essential." "If DDT is a devil," the SOS believed, then "it is a remarkably friendly devil."[41]

After CBS News aired a report on the release of the Mrak Commission report, the SOS sent letters of protest to CBS and members of Congress. From the SOS's perspective, CBS used the Mrak Commission report to support a false conclusion, telling its viewers that the report recommended a "complete ban on DDT uses." Instead, the Mrak Commission provided exemptions for use based on disease control and concern for "public welfare," if alternatives could not be found. The SOS claimed that, although still somewhat ambiguous, the recommendations found in the report, which advocated the suspension of certain uses of DDT, were ones the SOS's "members [could] certainly live with and endorse." Because of CBS's journalistic negligence, the SOS complained, supporters of DDT had "gotten short shrift." Betty Chapman, SOS's volunteer secretary, continued her admonishment of the CBS broadcast, "In science this is tragic, as the public can't distinguish between and among scientific arguments."[42]

In October 1971, the SOS received a letter of support from Norman Borlaug, who had received the Nobel Peace Prize the year before for his development of Green Revolution technologies. "Thank God some courageous volunteers are getting into the act of fighting back against the propaganda campaign against pesticides and fertilizers, which is based on emotion, mini-truths, maybe truths and downright falsehoods and launched by full-bellied philosophers, environmentalists and pseudo-ecologists," Borlaug wrote. "Unless this vicious, shortsighted propaganda campaign is stilled it will bring disaster to the world."

Not one to mince words, Borlaug had grown increasingly irate about the EDF legal campaign. "If such legislation is passed in the United States," Borlaug continued, "in developing countries it would be disastrous. Malaria will come back fast. The so-called Green Revolution in cereal crops will die." And all of this, Borlaug claimed, would "be the result of the combined shortsightedness, selfishness and egoism of a small group of privileged people."[43] Borlaug enclosed a $50 check to support the organization.

Despite its best efforts, the SOS had minimal success in promoting the successes of DDT to a wider audience. This failure was not for lack of effort; in many ways, the die had been cast. After years of dominating the public dialogue about the role of science in America, the chemical sciences found it increasing difficult to shape public opinion. Ecology, as both an idea and a scientific approach, resonated with many Americans frustrated with the unquestioning fanaticism and narrow vision of wreckers and exterminators.

Charles Wurster kept up his unceasing attack on DDT. As a biologist at the State University of New York, Stony Brook, Wurster published a number of articles on DDT's ecological impacts, including the rapid decline of the Long Island osprey population.[44] But it was his relentless opposition to DDT that led one author to describe him as "a Young Man with a Gripe."[45] Tenacious, hardheaded, and oftentimes belligerent, Wurster pursued his assault on DDT with unceasing vigor. "It is important to eliminate the use of DDT everywhere at the earliest possible moment," he wrote in a 1970 article. Wurster contended, "Mosquito resistance had diminished the effectiveness of DDT in malaria control programs. At the same time, however, the continued and excessive usage of DDT has hampered the development of safer control methods."[46] Yet Wurster also understood the possible benefits of DDT for public health work in dire cases. "Only in those malarious underdeveloped countries that for economic reasons can afford only DDT in their control programs," Wurster recommended, "does the continued use of DDT carry benefits that [remain] greater than the risks."[47] A year later, Wurster again made his case against the domestic use of DDT, while accepting its international use as a mosquito control agent. In an opinion essay published in the *New York Times*, Wurster told readers that the EDF had "not tried to cut off DDT from those underdeveloped countries." Wurster added, "Only there does the terrible human cost of malaria justify the additional environmental stress of further contamination by this destructive chemical, and even then [DDT] fails increasingly."[48] While Wurster and the EDF had grave reservations about the use of DDT in malaria control projects, believing the justified use of the chemical would be a slippery slope toward mosquito eradication, they were not against its use in the most pressing circumstances, nor was the SOS.

Nevertheless, Wurster, as the most public representative of the EDF, became the face of the anti-DDT campaign.

Not surprisingly, Wurster became a target of the chemical industry. And as it had in its attempt to paint Rachel Carson as an out-of-touch nature lover, the industry and its supporters cast Wurster as a radical ecologist out to destroy the American way of life. Wurster's politics seemed, to a great many Americans, downright dangerous.

Some of Wurster's (and the EDF's) most vociferous critics, expectedly, came from the scientific community and Congress. They included entomologist J. Gordon Edwards, biologist and former American Cyanamid scientist Thomas Jukes, and plant geneticist Norman Borlaug. Among congressional critics, Democratic representatives John Rarick of Louisiana, Jamie Whitten of Mississippi, and William R. Poage of Texas, who served as chair of the House Committee on Agriculture from 1967 to 1975, were the most outspoken. For his part, Rarick participated in one of the most controversial moments in the history of DDT, which took place at a hearing in 1971 before the House Committee on Agriculture.

The hearing focused on the safety of DDT and its effect on crop production, but what occurred would epitomize the contentious nature and interconnections of the DDT debate. For all intents and purposes, the hearing was a perfect storm—one that drew on the history of malaria eradication, agricultural development, the population crisis, and U.S. foreign aid, as well as the litigious domestic politics of the modern environmentalism. Readings or, perhaps more appropriately, misreadings of this event have had repercussions on environmental politics to this day.[49]

Edward Lee Rogers, the EDF's chief counsel, was called on to testify about the environmental impacts of DDT. His questioner was Rarick, who represented one of the nation's largest cotton-producing states. As a point of reference, Rarick called attention to a speech given to the Public Relations Luncheon Group of the Union League Club, New York, May 20, 1970, by former EDF lawyer Victor Yannacone, in which Yannacone claimed that his onetime colleague, Charles Wurster, suggested that the efforts to ban DDT would help solve the problems of overpopulation. "Was this the same Dr. Wurster," Representative Rarick asked, "whom Mr. Yannacone described as having said about the use of more toxic, yet less resistant pesticides, 'So what? People are the cause of all the problems. We have too many of them. We need to get rid of some of them and this is as good a way as any'"? After acknowledging that Wurster was indeed a founding member of the EDF, Rogers defended him, saying that the characterization of Wurster was highly inaccurate. Later, Rarick asked if this was the same Dr. Wurster "who said when asked the question, 'Doctor, how do you square this killing of people with the mere loss of some birds?' . . . [responded] 'It doesn't really make a lot of difference because the organo phosphate acts locally and kill [sic] farm workers and most of them are Mexicans and Negroes'"?[50]

J. Gordon Edwards later confirmed Yannacone's version of events, claiming, "[Yannacone] assured me Wurster *did* say that."[51] For advocates of DDT, this exchange confirmed their deep-seated suspicion about Wurster, the EDF, and, more broadly, the modern environmental movement: environmentalists cared more about trees then people; they were a relatively privileged group of Americans who were out of touch with the mainstream; and, at the heart of the matter, they were racists, despite their "liberal" politics.[52]

Wurster defended himself from these charges in a letter to the committee. "I wish to deny all the statements of Mr. Yannacone," he wrote, they "are pure fantasy and bear no resemblance to the truth."[53] The historical record neither confirms nor denies either version of the conversation. A thorough review of Wurster's papers at the EDF archives yielded no record of the above-mentioned statements, nor do they suggest that Wurster ever came close to describing farm workers or the problems of overpopulation in such terms. Additionally, an inquiry into the congressional archives also revealed little, as materials from the committee hearing could not be found, despite references to them in the *Congressional Record*. Notwithstanding the absence of a clearer picture of this episode, its significance cannot be understated. Not only did the exchange show Wurster in a negative light; it also cast a large shadow on the EDF and the environmental movement as a whole, a legacy that continues to shape the contemporary history of DDT.[54] A number of books, articles, and websites continue to cite Wurster's comments uncritically to advance an anti-environmental agenda.[55]

For supporters of DDT, bad news continued to arrive. Globally, malaria was on the rise. Reports from Ceylon (now Sri Lanka), the island nation off the southeast coast of India, indicated that nearly 1 million people contracted vivax malaria, even though Ceylon's eradication program was in the "consolidation phase."[56] Signs of malaria problems in India, Pakistan, and throughout Africa alarmed public health officials and government officials alike, suggesting that the promise of eradication had indeed come to an end. Within the United States, the publication and, perhaps more importantly, the popularity of Paul Ehrlich's *The Population Bomb* (1968) and Barry Commoner's *The Closing Circle: Nature, Man, and Technology* (1971) spread the gospel of ecology while underscoring the limits of American consumption. And much to the surprise of almost all Americans, there was an environmentalist in the White House.

On January 1, 1970, President Nixon signed into law one of the most sweeping pieces of environmental legislation ever crafted, the National Environmental Policy Act (NEPA). Broad in its conception, NEPA introduced a series of regulatory provisions aimed at reversing the myriad environmental problems the nation faced. Of particular importance was the requirement for all federal agencies to produce a detailed report, an environmental impact statement, assessing the environmental impacts of a proposed action. The legislation also created the Council on Environmental Quality, which would serve in an advisory capacity to the president. Upon signing the law, Nixon claimed that the "environmental decade" was at hand.[57]

Nixon had more surprises. Following the dramatic success of Earth Day on April 20, 1970, Nixon, quick to recognize the potential of this politically active constituency, established the Environmental Protection Agency (EPA) in July of that year, consolidating all environmental enforcement activities under one umbrella agency. To head up the EPA, Nixon called on William D. Ruckelshaus, a former assistant attorney general and onetime member of the Indiana House of Representatives. Despite the tremendous growth in environmental policymaking Nixon engendered, Ruckelshaus later remembered that the president "never asked me about anything going on in EPA. . . . He wasn't really curious about it."[58] Not surprisingly, Nixon's interest in the environment was purely political.

Nevertheless, NEPA and the creation of the EPA transformed the pesticide issue. Pursuant to Nixon's executive reorganization of environmental responsibilities, the registration and regulation of pesticides was transferred from the USDA to the EPA. Upon this transfer, the EDF petitioned Ruckelshaus twice, first asking him to review the secretary of agriculture's failure to suspend DDT, and second, requesting suspension of all uses of DDT during the process of cancellation, which, if approved, would prohibit any future use of the pesticide. As a result of these actions, a court ruling required Ruckelshaus to "issue immediately notices of cancellation of all uses of DDT."[59] He complied, stating, "The question is not whether the court's decisions are right as a matter of law, but rather the public's right to a full and open airing of a controversy surrounding the continued use of DDT."[60]

The court ruling led thirty-one DDT formulators and manufacturers to appeal the cancellation notice, calling for a public hearing under the law. The hearings that followed would last from August 1971 to March 1972 and would produce over 9,000 pages of testimony from over 125 witnesses (hereafter, these hearings will be referred to as the Consolidated DDT hearings). Unlike the Wisconsin DDT hearings, which placed the onus on the EDF to prove that DDT was a pollutant under the law, the Consolidated DDT hearings shifted the responsibility to the chemical industry. Nevertheless, the hearings relied on the same rules of evidence and cross-examination. The Consolidated DDT hearings would become the most significant ever held on DDT, and these hearings would hinge largely on the tension between the uses of DDT within the United States and beyond its boundaries.

On August 17, 1971, Hearing Examiner Edmund Sweeney opened the Consolidated DDT hearings by clarifying the issues at hand, which were based on three interrelated questions about labeling.[61] One was "whether the labeling accompanying the economic poison . . . contains directions for use which are necessary, and if complied with adequate for the protection of the public." The second involved "whether such label contains a warning or caution statement . . . [which,] if complied with, [is] adequate to prevent injury to living man and other vertebrate animals, vegetation and useful invertebrate animals." And last, whether DDT, "when used as directed or in accordance with commonly recognized practice, shall be injurious to living man or other vertebrate animals or vegetation except weeds to which it is applied or to the person applying such economic poison." Labeling requirements for pesticides date back to the original Federal Insecticide, Fungicide, and Rodenticide Act of 1947. Labels contained federal registration information, including manufacturers, types of uses, and information on the degree of toxicity. In its original form, FIFRA required the government to prove that pesticides were harmful. Later amendments shifted the burden of proof, prior to federal registration, onto the manufacturers.[62] The question before Examiner Sweeney was whether a label could be developed that would address the full extent of DDT's ecological and biological hazards. He concluded his summation of the issues by stating, "The burden in this hearing is on the petitioner registrants."[63]

Sweeney's selection as hearing examiner alarmed many environmentalists. Throughout the hearing, Sweeney displayed much disdain for the EDF, often limiting important testimony and even being openly hostile to EDF witnesses. In procedural matters of the law, Sweeney ruled frequently in favor of the chemical industry. On more than one occasion, a number of Fish and Wildlife Service scientists walked out of the hearing. Often exasperated, the EDF team felt more like the defense than the prosecution.[64]

On the whole, the Consolidated DDT hearings revolved around three main issues: the international use of DDT, its ecological impact, and cancer. Accordingly, the chemical industry's case evoked the three decades of success with DDT and the chemical's record of combating insect-borne disease, and it included a peculiar defense of DDT that relied on the fact that DDT use over the years had begun to wane to suggest that the chemical should not be banned until an alternative was found. In essence, a voluntary withdrawal of DDT proved that federal regulation was unnecessary.[65] Conversely, the interveners—the EPA and the EDF—presented evidence that tried to develop a distinction between the ecological well-being of the nation and public health needs abroad. Finally, the issue of cancer made these ecological and international questions much more difficult to answer. Throughout the hearing, there was no consensus on whether DDT was carcinogenic. Even today, these questions are difficult to answer. Studies strongly indicate that girls exposed to DDT at a young age have a rate of breast cancer occurrence five times that of women in general later in life.[66] Evidence also indicates the chemical is an endocrine disruptor, interfering with the sexual and reproductive functions of humans and wildlife.[67] But other studies on DDT's possible carcinogenic effects remain inconclusive. So while the issue of cancer complicated the hearings, it did not dissuade the participants from engaging in a lengthy debate about the ecological hazards and human health benefits of the chemical.

Indeed, the chemical industry's legal counsel, Robert L. Ackerly, initiated the proceeding by suggesting that the case could not "be decided in a vacuum." Ackerly concluded, "Thus it is necessary to understand the history of DDT use in this country and worldwide."[68] Ackerly's history lesson involved details of the WHO's malaria eradication program, which he characterized as having "achieved [eradication] in areas populated by some 700 million people, [while] active control programs . . . were sharply reducing exposure for over 1 billion additional people."[69] Ackerly then presented witnesses who spoke glowingly of DDT's success in combating malaria, including Edward Knipling (former director of the USDA's Agricultural Research Service), Norman Borlaug ("father" of the Green Revolution), and Surgeon General Jesse L. Steinfeld, a cancer specialist and advocate of comprehensive public health services.

Knipling and Borlaug brought enormous scientific prestige to the hearing. Although he had recently stepped down as director of the USDA's entomology research division, Knipling was widely recognized for his contribution to entomology and agricultural science, having received the Rockefeller Prize for Public Service and the National Medal of Science, as well as having been named one of the "100 most important people in the world" by *Esquire* magazine in 1970. He remained steadfast in his belief that farmers should have a full arsenal of insect control methods at their disposal, including chemical and biological controls. Because of his sterilization work on screwworms, some scientists argued that Knipling had developed "the single most original thought in the 20th century," while others called him the "father of male sterilization technique."[70]

Indeed, his interest in and commitment to biological control made him a forceful, and perhaps reasonable, voice in the DDT debate. "I think we were a little late in recognizing its problems," Knipling told the *New York Times Magazine*, "and I think we were a bit shortsighted in putting nearly all our effort in chemicals. But without DDT a lot of money and crops would have been lost. A lot of people would have died from malaria and other diseases. Yet we now realize that DDT is not the final answer."[71]

In contrast, Norman Borlaug showed little respect for environmental organizations like the EDF and the politics they engendered. Borlaug's authority as a scientific voice of reason and a Nobel Peace Prize recipient resonated with a great number of people, especially those who believed in the transformative power of technology. In "Where Ecologists Go Wrong," Borlaug argued, "If agriculture is denied [pesticides'] use because of unwise legislation that is now being promoted by a powerful lobby group of hysterical environmentalists, who are provoking fear by predicting doom for the world by chemical poisoning, then the world will be doomed—but not by chemical poisoning, but from starvation."[72] Ecologists certainly bore the brunt of Borlaug's criticism, being viewed as a group with only "superficial expertise."[73] Yet his approach to ecology also reflected his disinterest in other forms of scientific knowledge, whether they were ecological science or local knowledge of food production. His belief in the superiority of technological science blinded him to other forms of knowledge.

Knipling's testimony raised the specter of the effects a domestic ban on DDT would have on global consumption. He indicated, "The suspension of an insecticide such as DDT for all purposes would have an impact on the thinking and the feeding of people in other countries where they might need the material for health purposes to a greater extent than we, where the need to produce food in adequate numbers and cheaply and safely from the standpoint of the food itself are far greater than they are in the United States."[74]

Knipling reaffirmed his position on cross-examination. EDF attorney William Butler asked Knipling, "You suggested that it was of concern to you what developing countries with malaria problems would do if uses of DDT were cancelled in this country. . . . Doesn't the effect on developing countries of

U.S. restrictions on DDT depend on the reasons this country gives for its restrictions on DDT?" Knipling responded, "I would think so, largely, but I do know the psychological effect it has. I will say this, that it is unfortunate in a way; the reasons why I am strongly for eventually minimizing the use of DDT or getting rid of it completely are environmental and not a hazard to people directly. But I think that 90 percent of the people think that DDT is a dangerous insecticide to even use in the house, which is not true. It is one of the safest insecticides we've ever had."[75]

As a champion of agricultural development and modernization projects, Borlaug, in his appearance at the hearings, puzzled many from the EDF, while causing excitement in others, namely Examiner Sweeney, who seemed delighted to have a Nobel Prize recipient in his court. Wurster commented, "Borlaug should not have been permitted . . . to testify where only expert testimony was properly admissible."[76] In a letter to an EDF supporter, Wurster again called into question Borlaug's expertise, stating that a chemical industry scientist "could hardly have shown more clearly how little he knows about the DDT issue than by choosing such a bedfellow."[77]

During his testimony, Borlaug spoke about his past work in Mexico, about the process of increasing wheat and corn yields in developing nations, and about how U.S. knowledge and capital served to underwrite the Green Revolution. Calling on his experience of twenty-seven years abroad, he commented that a "great misunderstanding . . . is going on in the world" and that he was "concerned" for people in developing nations "because of some of the things that [he saw] on the horizon" in the United States. Borlaug decried the "oversimplification of this whole question of balance in nature."[78] Asked what the ramifications of a domestic ban would be in Mexico, Borlaug responded, "I worked these 27 years outside. I know the psychology of the peoples and especially as it relates to public health, it would be most unfortunate . . . in terms of public health—I don't like to use the term, but—disastrous."[79] Borlaug continued defending the use of DDT abroad under direct examination and cross-examination, restating his opinion that a domestic ban would have severe consequences overseas. At the conclusion of his testimony, Borlaug held a press conference to reaffirm his testimony. Montrose Chemical Corporation, the lone domestic manufacturer of DDT, arranged the press conference.[80]

To support this position, the chemical industry called on Surgeon General Jesse Steinfeld to testify. Reading from a prepared statement, Steinfeld appeared before the examiner "as an advocate for the continued availability of DDT in the public health armamentarium—not only in this country but in the rest of the world."[81] Steinfeld continued, "The United States is the principal source of the DDT used in malaria eradication. The U.S. produced DDT water dispersible powder (WDP) is the highest quality and the most economical formulation in the world. . . . If the U.S. Government policy toward the use of DDT were that drastic, and this is the question whether or not AID [United States Agency for International Development] would permit the continued use of U.S. funds for the purchase of DDT from non-U.S. sources. Therefore, a decision that would result in stopping the production of DDT in the United States could, in essence, be a denial of the use of DDT to some of the most highly malarious areas of the world." The suspension of DDT for being "an imminent hazard to the public," Steinfeld contended, would have "deleterious repercussions for the malaria eradication efforts by raising unwarranted fears in the minds of those responsible for decision making in other governments." "However," Steinfeld asserted, DDT's "continued production and availability . . . would leave the final decisions on the use of DDT in each malarious country where it should rest—in the hands of the policymakers in each of the countries."[82] Steinfeld's observations reflected the impact of DDT on malaria eradication projects and called attention to the "quality" of U.S.-made DDT, suggesting the American production of the chemical was vastly superior and the most cost-effective product on the market. Perhaps more importantly, Steinfeld understood DDT simply as an instrument for public health, even though it might have been an "imminent hazard" to U.S. citizens.

Butler's cross-examination explored many of these issues. He inquired if the differences in wealth between the United States and developing nations allowed for the use of more costly pesticides within the United States, pesticides such as malathion. Steinfeld responded to this line of questioning by "pleading for . . . the option to have [DDT] available for the purpose of protecting the public health . . . cost not being an issue in the United States."[83] Butler asked about the "psychological effect" a domestic ban would have on other countries, and Steinfeld reiterated his belief that "if the United States banned or stopped the use of a particular compound because of its possible carcinogenicity, mutagenicity, teratogenicity, or some other terrible things, certain individuals who make decisions who may not be scientists or physicians in other countries might undertake actions which would not be as rational as we would hope they would be."[84]

While the case for the petitioners and the USDA rested squarely on DDT's safety and its use in malaria eradication, the EPA and the EDF made a complex presentation that was built on the successful campaign in Wisconsin.

In his opening statement, however, Butler attempted to define the political boundaries of the hearing, stating, "The geographical scope of what we feel is at issue here is the United States, although there may be, in any given instance, relevant examples drawn from a foreign experience."[85] Butler offered evidence related to the rapid decline in the raptor population. Expert witnesses pointed to the hazards of DDT to fish and other marine life. Others testified to DDT's possible carcinogenic qualities. Again, based on its experience in Wisconsin, the EDF, along with the EPA, relied heavily on a multidisciplinary approach to the question of DDT's danger to the environment and to people.

Along with this critical testimony about DDT's hazards, the EDF also presented witnesses who offered alternatives to chemical agriculture. Robert van den Bosch, a biologist and critic of the pesticide industry who had testified in Wisconsin, reappeared on behalf of the EDF to submit critical evidence supporting the concept of integrated pest control, a method that combines biological techniques with the selective use of nonpersistent insecticides. Part of the EDF strategy throughout the DDT controversy was to argue that within the public health and agricultural realm, alternatives to pest controls did exist that were less environmentally harmful and less costly. In promoting an integrated pest management solution, the EDF developed the centerpiece of its institutional mission, offering solutions to critical problems, not simply taking legal action against a perceived environmental problem.

The testimony presented to Examiner Sweeney crossed many disciplinary boundaries—drawing on ecology, biochemistry, agronomy, toxicology—and it also crossed many political borders—encompassing Africa, Southeast Asia, India, and numerous states within the United States. By crossing these borders, the chemical industry wanted to retain its right to market and manufacture DDT domestically and sell it globally. The EDF, on the other hand, emphasized the fluidity of borders to argue that DDT was a worldwide contaminant, but suggested that the nation's borders established the jurisdictional boundaries of the hearing.

The final arguments reaffirmed these ideas about borders and boundaries. USDA attorney Elliot Metcalfe reflected on the nine months of testimony, arguing that the international issues involved in the case created a problem. He stated, "It is manifest in Respondent's analysis that they [the EPA and EDF] have taken two diametrically opposed positions. On the one hand, they allege that DDT is the cause of cancer and genetic defects in man and, in the same instance, they say that DDT should be available for the Public Health Service here in the United States and . . . acknowledge that it should be used overseas in foreign nations. It's quite absurd to think that Respondent's position would be that we should subject foreign citizens, and citizens in this country, to a compound which they classify as a mutagen and a carcinogen."[86] For Butler and the EDF, the legal argument was rather straightforward. Butler simply stated, "There may be some chemicals—DDT being one—that no label can be written for, because it simply cannot be used in the terms which are required in FIFRA."[87]

Despite the EDF's reading of the regulation and their evidentiary presentation, Examiner Sweeney ruled in favor of the chemical industry. Examiner Sweeney argued, the "USDA and Group-Petitioners have met their burden of proving that the labels involved here meet all the requirements of the act . . . [and] that the uses of DDT under their registrations do not create an unreasonable risk on balance with the benefits."[88] In his findings, Examiner Sweeney addressed the presentation of the global use of DDT and the jurisdictional limits of the hearing, stating that "such limitless ambit of testimony possibly could be pertinent in a rule-making proceeding. . . . But the fact is: this hearing is not a rule-making proceeding. This hearing is an adjudicatory proceeding in every sense of the word."[89]

Although much of the evidence presented at the hearing was drawn from beyond U.S. borders, with the petitioner's case resting largely on the overwhelming success of DDT overseas, Sweeney's ruling upheld a narrow interpretation of U.S. law. His findings did not take into account the testimony about the global reach of DDT or the success of malaria eradication programs. In fact, Sweeney stated, "The purpose of expending so much expert opinion to make that point [on the global presence of DDT] is somewhat obscure to me."[90]

For those following the hearings, the decision was surprising, but not a surprise. Sweeney's obvious preference for the industry's position all but made the ruling a foregone conclusion. Yet the decision did elicit condemnation.

A swirl of controversy erupted at the release of Sweeney's ruling. The *New York Times* called Sweeney an "apologist for DDT," claiming his report "flies in the face . . . of the testimony [and] flatly contradicts the opinion of scientists."[91] The EDF called the examiner's ruling "worthless" and said it "should be disregarded."[92] Even prior to the official release of Sweeney's ruling, the EDF filed exceptions to the report, challenging not only the findings, but also the reading of the law and Sweeney's evidentiary rulings during the hearing. In its challenges, the EDF reiterated the narrow scope of the hearing: "The Examiner states that there are two areas of contention at this hearing: 'the global benefits and risks of DDT for all uses; and the particular benefits and risks of specific uses of DDT, such as on an agricultural product.' Only the question of the 'benefits and risks from remaining domestic uses of DDT is at issue,' as the Examiner paradoxically admits four pages latter."

Sweeney's decision, however, was not the final say. William Ruckelshaus was charged with the final determination on DDT. On June 14, Ruckelshaus released his final ruling, overturning Sweeney's findings and concluding that the "long-range risks of continued use of DDT . . . [were] unacceptable and outweigh[ed] any benefits." To many, the ban represented the triumph of modern environmentalism and the fulfillment of liberal environmental governance. Yet it did not bring an end to the debate. Rather, the ruling only strengthened the resolve of all participants to continue advocating the use or suspension of DDT and other chemical pesticides. Moreover, by casting the question of DDT as primarily a legal matter, the EDF chose a decision-making process that, at its root, was adversarial. The process could yield only one winner, so the decision reverberated longer than the logic behind it. For example, shortly after the decision, J. Gordon Edwards issued a press release, "The Infamous Ruckelshaus DDT Decision," that was subsequently entered into the *Congressional Record* by Senator Barry Goldwater. Edwards believed the decision was "an abject capitulation to professional environmental extremists and a tremendous defeat for science and mankind."[93]

Certainly, the divisions that arose from the Wisconsin hearing and that were further developed throughout the Consolidated DDT hearings limited the possibility of engaging in a meaningful debate about the implications of DDT use for international public health.[94] The divisiveness of these debates was not lost on the American public. In a letter to William Poage, one concerned citizen wrote, "Many claim starvation if chemicals are banned! But none has done nature justice of seriously trying to reduce chemical use by biological controls. For them it's all or nothing."[95] Another wrote in response to a series of articles published in *Smithsonian* that tried to weigh "both sides" of the issue: "The arguments miss each others; your 'pro' is arguing almost exclusively for continuing use in poor countries, while your 'contra' is arguing against continuing use in the United States."[96]

In making his judgment, Ruckelshaus weighed the evidence based on the domestic risks and benefits, noting, "These hearings have never involved the use of DDT by other nations in their health control programs."[97] However, Ruckelshaus's ruling allowed for the continued manufacture and export of DDT. There was no cancellation order. Some of that production, in Ruckelshaus's opinion, should be used for potential "health and quarantine" programs administered by "U.S. Government officials and State Health Departments who will be knowledgeable as to the most effective means of control and mindful of the risks of using DDT."[98] The rest, supposedly, would be available for export.

So despite the domestic ban, the ruling alluded to a much more nuanced interpretation of the issue, an interpretation that was closer in design to the Mrak Commission report than either the EDF or the chemical industry assumed. Yet the contentious nature of the debate only hardened attitudes about the meaning of the decision, leaving both sides more committed to advancing their views about technology and ecology. Many of these ideas would be played out throughout the decade, leading to an even more complex set of questions as the American Century began to wane.

The production of DDT and the export of a banned substance created a paradoxical set of issues that further muddied the boundaries of environmental regulation. Upon learning of Ruckelshaus's ruling, Congressman George A. Goodling (R-Pa.) bluntly stated, "It may be a bit ironic that DDT is banned for use in the United States but its manufacture and sale to foreign countries has the blessing of EPA. There is a possibility that it could be sprayed on food and fiber plants and in turn the end product exported to the United States. Hardly appears consistent, does it?"[99] The search for regulatory consistency is what ultimately drove a wedge between advocates and opponents of DDT. Either DDT was an invaluable pesticide, or it was not; there was no in-between. In the struggle to make a consistent argument on the DDT issue, however, the chemical industry and the EDF neglected the merits of each other's position. The Consolidated DDT hearings also demonstrated that both sides of the debate understood the world beyond the U.S. borders as simply an extension of American interests. Whether they were promoting the strategic designs of technological modernity or the view that ecology binds the entire world together, participants in the contentious debate over DDT shared a similar vision—what is good for the United States is good for the rest of the world.

The inconsistency of the Ruckelshaus ruling, moreover, set the stage for a new chapter in DDT's contentious history, one that challenged the global reach of the United States, the limits of environmental regulation, and the meaning of DDT and the American Century.

ONE MAN'S PESTICIDE IS ANOTHER MAN'S POISON
The Controversy Continues

It's unrealistic to believe that there will be a solution
through technological breakthrough.
—J. George Harrar, Rockefeller Foundation president,
before a hearing on world food and population
problems, 1966

I do know that [the EPA] has established a considerable record
for doing the unreasonable and the unnecessary.
—Representative William R. Poage, 1974

In 1970, William R. "Bob" Poage, the representative from the 11th Congressional District in Texas and chairman of the House Agriculture Committee, introduced a new bill for consideration. Poage, who a few years earlier had traveled extensively in India and Africa on a congressional junket, was acutely aware of the interconnections between the U.S. environment and the world. During the late 1960s, he was firmly committed to Johnson's Food for Peace program, a project designed to use American surplus grain to feed the hungry in India. Poage also represented a congressional district of farmers. The agricultural plains of central Texas hardened Poage's constituency to the reality of growing crops in harsh environments, and Poage became a forceful advocate on behalf of those constituents. As Agriculture Committee chair, Poage also understood the complicated and costly decisions farmers continuously made to ensure crops headed to market. Across the nation, rising production costs and increased capital investment made agricultural work risky. This risk increased precipitously over the previous two decades as agricultural chemicals became the most accepted form of pest control in the nation. With nearly 33,000 registered chemicals on the market, the potential chemical combinations of insect and weed control only amplified the risk. Yet increased federal regulation designed to mitigate the risk to human bodies and the environment had the potential to add to the financial risks for Poage's constituency.

Poage quickly grasped the domestic and international implications of increased federal interest in the pesticide problem. According to Poage, new regulations being proposed by the Nixon administration would put American farmers at a competitive disadvantage in the world market. He therefore proposed a resolution that would prohibit the importation of any agricultural commodity unless the producing country had restrictions on pesticides that were "at least equal" to those in the United States. Poage claimed that his legislation would both protect American producers from unfair foreign competition and shield American consumers from the potential hazards of chemical residues on imported goods.[1] "If residues of herbicides and pesticides applied here in the United States are harmful to American consumers," Poage speculated, "then certainly the same residues which are contained in agricultural commodities imported into the United States must be just as harmful. I don't think we should play any favorites, if in fact such a danger to consumers exists."[2]

Needless to say, the proposed bill ruffled the feathers of many agricultural importers. A rash of letters, many from major trade associations, flooded Poage's office, expressing concern about the proposed legislation. The Chocolate Manufacturers Association proclaimed its outrage over the bill because all cocoa beans had to be "imported from abroad" since they could not be "grown in the United States." Likewise, the Green Coffee Association of New York City and the National Coffee Association sent letters pointing out "certain grave problems" with the bill. Standard Fruit argued that the bill would be "particularly undiplomatic at the present time." "Such action," Standard Fruit claimed, "would be viewed by foreign governments as an invasion of their sovereignty, and bitterly resented."[3] This was a rather remarkable statement from a company with a long history of interventionist policies.[4] While such indignation could be expected from agricultural trade associations, since they would be directly affected by the legislation, the nation's leading producer of baby food, Gerber, also expressed dismay at the pending legislation, although Gerber, unlike agricultural commodity producers, worried more about potential trade restrictions than about pesticide residues. Calling the bill "protectionist in nature," Gerber claimed that it could encourage other nations "to erect similar non-tariff barriers against U.S. agricultural exports."[5]

Poage had indeed raised a particularly complex issue that in many ways would set the stage for a series of debates over the international implications of domestic pesticide regulation in the years to come. While it is difficult to discern Poage's true intent regarding his proposed legislation, demonstrating the hypocrisy of federal policy would be high on the list, since he would become one of the most outspoken critics of federal environmental regulation. More importantly, the bill confirmed that the politics of pesticides was quickly becoming a national *and* international issue. In an effort to provide a less confrontational approach to the pesticide issue, Assistant Secretary of Agriculture Richard Lyng argued, "The solution to the worldwide concerns about health and environmental matters does not lie in unilateral action by any nation against other nations. It lies in concerted international action . . . [and] through accelerated international cooperation."[6]

Poage's legislative action, similar to actions debated during the Consolidated DDT hearings, attempted to assess the risks of DDT on both sides of the border. While the Poage bill ultimately failed because of its protectionist point of view, the Consolidated DDT hearings reinforced the concept that the United States was "a nation among nations."[7] Although much of the testimony was highly contentious, one issue remained clear to all participants: DDT was a global chemical. Not only was it sprayed throughout the world, but since it was a persistent pesticide, it migrated through soils and across rivers, lakes, and oceans. DDT also migrated across borders as residue on agricultural commodities. But the global use of DDT also saved lives. Despite these contradictions, what was apparent is that political borders could not control the long-term impact of the pesticide, leaving American policymakers with a complex environmental and regulatory problem.

TWO ORGANIZATIONS SHAPED the U.S. environmental agenda: the Council on Environmental Quality (CEQ) and the Environmental Protection Agency (EPA). Upon assuming the CEQ chairmanship, Russell Train created an organization in his image. A lawyer, naturalist, and diplomat, the energetic Train understood that the nation's environmental problems did not stop at the border. During his tenure, he took great interest in international affairs, traveling the world and speaking with leaders about an array of environmental issues.[8] He encouraged bilateral talks with the Soviet Union and China about the environment. But, perhaps most of all, he liked to spread the gospel about the U.S. commitment to saving the environment at home and abroad. Train boasted that in the early 1970s the United States was "the clear world leader ... in terms of both domestic environmental policy and international cooperation."[9] Since the Second World War, the American Century was a largely technological project, predicated on U.S. leadership on the global stage. Train brought an ecological perspective to this project.

William Ruckelshaus, the EPA administrator, remained focused on domestic problems. And he had his plate full. Not only were the administrative burdens of consolidating over 5,000 employees from various federal agencies and departments under one roof difficult, but the nation's environmental problems were incredibly complex.[10] Water pollution despoiled the nation's lakes, rivers, and streams. Smog and other forms of air pollution clouded much of the United States. Toxic chemicals quietly seeped through soils and flowed through the bloodstreams of all human and animal life. It seemed that nature had had enough of American postwar prosperity. For his part, Ruckelshaus established an aggressive agenda to solve some of the nation's most pressing problems, namely air quality, water pollution, and pesticides. The Consolidated DDT hearings were set against this vastly changing regulatory landscape.

Though the Ruckelshaus DDT decision was controversial—prohibiting the chemical's domestic use but allowing domestic manufacture and export to continue—it was remarkable for another reason. Rather than make the pronouncement from the halls of Washington, which would have symbolized the growing strength of federal environmental governance, Ruckelshaus made his announcement in Stockholm, Sweden, against the backdrop of the first United Nations conference on the human environment. The location selected for the press conference had much to do with the theater of environmental politics and the changing context in which environmental decisions were made.

If 1970 marks the beginning of the "environmental decade," then 1972 represents the birth of a global environmental movement. At the suggestion of Swedish officials, the United Nations sponsored the international conference on the human environment. Representatives from 114 nations, 16 intergovernmental agencies, and over 250 nongovernmental organizations converged in Stockholm to discuss a broad array of environmental issues confronting developed and developing nations disproportionately. At issue were the environmental problems caused by development (i.e., air and water pollution, industrial waste, etc.) and the promise of development to ease environmental problems in the "developing word" (i.e., poverty, famine, and disease).[11] In the unofficial report from the Stockholm Conference, *Only One Earth*, British economist Barbara Ward and French biologist René Dubos, who is notable for coining the expression, "Think Globally, Act Locally," wrote, "There was general agreement among [attendees] that environmental problems are becoming increasingly world-wide and therefore demand a global approach."[12] According to U.N. Under–Secretary General Maurice Strong, "Stockholm launched a new liberation movement—liberation from man's thralldom to the new destructive forces which he, himself, has created."[13] For Strong and other conference participants, the time was at hand to begin the process of healing the planet.

Drawing attention to the nation's commitment to international environmental governance, Ruckelshaus and Train were part of a thirty-five-person U.S. contingent. Other participants included Christian A. Herter Jr., deputy assistant secretary of state for environmental and population affairs; Rogers C. B. Morton, secretary of the interior; Laurence Rockefeller, chairman of the Citizens Advisory Committee on Environmental Quality; and former movie star Shirley Temple Black, who was a U.S. delegate to the United Nations. The size of the U.S. delegation suggested the overwhelming U.S. commitment to global environmental affairs. But more importantly, its makeup highlighted the ways in which environmental issues had become part of both domestic and foreign policy. Interagency cooperation and conflict, however, would continually shape this new environmental ethos and pesticide policy for the foreseeable future. The problem with DDT would only highlight these issues.

Despite the national and international efforts to identify and regulate the flow of chemicals across borders, American chemical producers continued to market their products overseas. For U.S.-based Montrose Chemical Corporation, the EPA ruling meant increased profits.

Beginning in 1947, Montrose Chemical produced technical-grade DDT at its facility in Torrance, California. At the height of its production, from 1968 to 1970, Montrose Chemical turned out nearly 80 million pounds of the pesticide, making Montrose the nation's leading DDT producer.[14] By the early 1970s, Montrose Chemical held a virtual monopoly on the domestic production of DDT. Other producers, Olin Mathieson, Allied Chemical, Diamond Shamrock, and Lebanon Chemical, abandoned the DDT market altogether in the wake of new EPA regulations and shrinking profit margins.[15] Montrose, on the other hand, established itself firmly in both the domestic and the international marketplaces and, for a time, was able not only to weather the environmental regulation, but to prosper. A major factor in the company's success was that the quality of the DDT Montrose produced far exceeded that of their competitors, making the Montrose brand a more effective and valued insecticide. Montrose became the last domestic manufacturer of technical-grade DDT, making the early 1970s a highly profitable period for the company. Montrose president Samuel Rotrosen later remembered, "The [EPA] decision did not impact us in that it made us stop making DDT. As a matter of fact, we had the most profitable years of our corporate history in the few years following the decision."[16]

Similarly, Stauffer Chemical Company, which had a 50 percent controlling interest in Montrose, embarked on a ten-year period of tremendous growth. Total sales doubled from 1972 to 1976, from $543 million to $1.1 billion. During that same period, sales of Stauffer's agricultural chemical line rose from $70 million to $198 million, and increased to $299 million by 1979.[17] Likewise, revenue from Stauffer's international sales division increased from $167 million in 1975 to $271 million in 1979.[18] Other chemical corporations followed suit. In 1977, DuPont garnered almost one-quarter of its total sales from the international market. Dow, Union Carbide, and Monsanto reported similar sales figures for foreign sales.[19] Between 1947 and 1978, annual domestic chemical production rose more than 900 percent.[20] In 1975, the International Trade Commission reported that production of pesticides doubled to almost 1.6 billion pounds during the previous four years.[21]

But U.S. manufacturers were not the sole suppliers of chemical pesticides in the world market. European chemical producers played a critical role in internationalizing pesticides as well, producers including Ciba-Geigy (Switzerland), ICI (England), Bayer (Germany), and Rhône-Poulenc (France), which were the four largest pesticide corporations in the world by the end of the 1980s.[22] Indeed, one of the paradoxes of the "environmental decade" was the incredible growth of the pesticide industry. The chemical industry made inroads into almost every national market around the world. Because of the strength of European industries and the lack of export regulations in Switzerland, France, and Great Britain, U.S. manufacturers believed that federal regulation of the export market would put them at a competitive disadvantage.[23]

In 1972, Congress amended FIFRA with the passage of the Federal Environmental Pesticide Control Act (FEPCA), tightening regulation on domestic pesticide use.[24] The new law placed additional regulatory burdens on chemical interests, added new labeling requirements, established record-keeping procedures for chemical manufacturers, allowed the EPA to seize illegal pesticides, and permitted registration only if it could be determined that the pesticide would not cause unreasonable adverse effects on the environment.

FEPCA supplanted state efforts to regulate pesticides, even those chemical compounds that did not cross state lines.

Chemical companies simply sidestepped this by exporting pesticides without federal registration. FEPCA allowed certain environmentally dangerous pesticides to receive federal permits despite the more stringent regulations, while unregistered pesticides for export were not subject to EPA review. Because of congressional resistance, existing loopholes, and the upcoming presidential election, environmentalists were quite disappointed with the outcome of the regulation.[25] The *New York Times* made this point quite clearly. Alluding to Theodore Roosevelt's conservation and foreign policy legacy, the newspaper announced, "Once again the Nixon Administration is displaying its tendency in environmental matters to talk boldly and carry a small twig."[26]

Nevertheless, FEPCA imposed a new regulatory burden on domestic chemical manufacturers. Global competition, increased regulatory and production costs, and the continued cost of research and development left many American producers strategizing how to navigate this new regulatory terrain and global marketplace. Samuel Rotrosen claimed, "Montrose was in the business of selling DDT and we looked for markets wherever we could find them."[27] Dow Chemical chairman Carl A. Gerstacker made perhaps the most provocative statement regarding federal regulation and the global chemical market. Speaking at the White House conference "The Industrial World Ahead" in February 1972, Gerstacker stated: "I have long dreamed of buying an island owned by no nation . . . and of establishing the World Headquarters of the Dow company on the truly neutral ground of such an island, beholden to no nation or society. If we were located on such truly neutral ground we could then really operate in the United States as U.S. citizens, in Japan as Japanese citizens and in Brazil as Brazilians rather than being governed in prime by the laws of the United States. . . . We could even pay any natives handsomely to move elsewhere."[28]

Gerstacker's dream of fluid borders and shifting corporate identities, flowing as freely as chemical agents across national boundaries, underscores the profound changes in both the global environment and the global marketplace. Without question, chemicals were rapidly becoming part of the total environment. They were, in fact, inescapable. But within this increasingly toxic environment, political boundaries also created specific differences. Each nation held its own ideas about how to protect, consume, or preserve natural environments. The number of countries with national environmental agencies increased from 26 to 144 between 1972 and 1982, making these national distinctions that much clearer.[29] However, the legacy of conquest, colonialism, and capitalism often impinged on the ability of some nations to regulate their natural environments. Indeed, Gerstacker's reference to relocating "natives" evokes a long and contested history of capitalism, resource exploitation, and racism that was lost on many American corporations. While Gerstacker's dream went unfulfilled (there is no Dow Island), in the context of understanding the new regulatory landscape, the desire to transcend boundaries as a way of circumventing environmental regulation and the impulse to shed a specific national identity portended the rapid shift in the global economy.

By 1973, the regulatory climate had changed dramatically as well. Nixon was consumed by the Watergate crisis, not that he paid much attention to environmental issues anyway. But perhaps more significantly, the turning point of 1973 came with a change in leadership at the EPA. Train stepped into the EPA administrator position when Nixon reassigned Ruckelshaus to serve as acting director of the FBI.[30] Train commented on the shifting political landscape during the early years of the EPA. "The environmental honeymoon," he wrote, "had come to an end."[31]

Traversing the new political terrain proved difficult for Train and the EPA. The looming energy crisis cast a long shadow over environmental politics, reflecting the nation's overreliance on foreign oil and the impact of environmental regulations that, in the judgment of many business leaders, placed undue hardships on American industry.[32] "As the economy soured," Train remarked, "environmental programs . . . became the whipping boy for inflation and job losses."[33] The shift from a period of rulemaking to enforcement marked a dramatic change for the EPA, with the energy crisis only adding to the problems of environmental enforcement.

Deindustrialization and the immediate threat of energy restrictions underscored the limits of American environmental politics. The massive demographic shift, in which a generation of Americans moved to the regulation-free regions of the American South and West, further imprinted an anti-Washington ethos on environmental policy.

More broadly, Americans grew disillusioned with the direction in which the country was going. In terms of domestic policy, the nation was adrift. And questions about the direction of U.S. foreign policy arose as the heated rhetoric of the Cold War cooled with the vague policy ambitions of détente. Increasingly, the divisions between right and wrong seemed to dissolve into a morass of confusion and chaos. To many Americans, the enormous structural changes brought about by the radicalism of the 1960s, as well as the corporate strategies of the 1970s that shifted manufacturing jobs overseas, represented a bitter and confusing end to American power.

Perhaps more symbolically, the recorder of the American Century, *Life*, ended its historic publication run in 1972. No longer did *Life*'s photographs and exposés on the American middle class, or its clearly defined narratives about U.S. power, capture the public's attention. Television instead conveyed the tumult of the 1960s and 1970s, and conveyed it visually. And while Henry Luce's magazine empire continued well after his death in 1967, the vision he so clearly articulated thirty years before seemed to have vanished. *Life* had come to an end. The question on the minds of many Americans was whether the American Century had faded away as well.

Debates over DDT, and pesticides more generally, symbolized these larger structural processes, which also reflected deep-seated anxieties about the changing world and the place of the United States in it.

Despite these massive structural and political changes, the EDF continued to press the EPA to control environmentally dangerous pesticides. In 1974, as a result of another EDF lawsuit, one followed by a lengthy judicial hearing, the EPA banned aldrin and dieldrin, two chlorinated hydrocarbon compounds with persistent qualities similar those of DDT. When EPA administrator Train issued the cancellation order, he also suspended production because aldrin and dieldrin were found to be an "imminent hazard" to public health.[34] Laboratory research determined that dieldrin and aldrin produced tumors in mice and rats. Dieldrin also was not a miracle compound for global insect eradication, despite being used during the WHO's malaria eradication campaign. Because of its carcinogenicity and limited public health value, Train suspended the use and production of dieldrin in the United States.

A year later, Train ruled against Velsicol Chemical Corporation, manufacturer of heptachlor and chlordane (two other chlorinated hydrocarbons), when he ordered an end to production because studies had shown the chemicals produced cancers in laboratory animals.[35] Federal regulation ended the domestic health threat associated with the use and production of these pesticides; however, heptachlor and chlordane continued to pose enormous risks overseas. In response to Train's ruling, corporations simply moved production and research facilities abroad. The National Pest Control Association claimed, "EPA action in registration and enforcement policy have discouraged continued pesticide research and development in the United States. One major pesticide producer is alleged to have said that in this atmosphere,

40% of its R&D will be conducted abroad by 1980."[36] In another case, Shell moved production facilities from California to the Netherlands and continued to manufacture aldrin.[37]

This latest cancellation order raised the ire of many people. A representative from the General Exterminating Company declared, "Stop Russell Train NOW!!" An angry citizen accused the EPA of "dictorial [sic] . . . tactics." Another critic of Train's leadership urged Congress "to repeal all environmental laws" on the grounds that they had "done nothing for the environment" and had worked to transform the United States into a "socialist dictatorship." The outraged letter writer concluded, "The Environmental Protection Agency must be abolished." Infuriated with Train's decision, Bob Poage introduced legislation to limit the EPA's authority to cancel pesticides unequivocally.[38]

The Poage-Wampler amendment aimed to give the USDA veto power over EPA cancellation notices. Although defeated by a mere eight votes in the House, the Poage-Wampler amendment expressed mounting opposition to the regulatory authority of the EPA. "Mr. Poage," one legislative aid wrote, "is a strong believer in maintaining environmental quality as well as banning those pesticides which are known for sure to be detrimental to human health. However, Mr. Poage is also in favor of feeding the 225 million people who inhabit the United States as well as other peoples of the world." Another letter from Poage's congressional office made this point much more directly: "Although Mr. Poage is concerned about our environment, he is more concerned with the ability of our nation's agriculture to produce food for our people. I doubt that very few people will give much thought to a clean environment if they are about to starve to death. . . . With the current world population growth rate it is not inconceivable that mass famine will occur in several parts of the world during our generation."[39]

Global population, food production, and chemical pesticides were never far from the minds of U.S. policymakers. For antiregulatory crusaders like Poage, international concerns over food and public health were useful in their argument against environmental regulation. But for many foreign policy experts, increasing crop yields was paramount in ensuring the safety and well-being of millions of malnourished people. And increased crop yields benefited America's standing abroad. Speaking at the World Food Conference in 1974, Secretary of State Henry Kissinger reiterated this belief, proclaiming, "We are convinced that the world faces a challenge new in its severity, its pervasiveness, and its global dimension." Drawing on ecological principles that would have astonished Rachel Carson, Kissinger told his audience, "We are faced not just with the problem of food but with the accelerating momentum of our interdependence." Indeed, it became increasingly apparent to most Americans and foreign policy experts that the problems of the world could not be solved by individual nations. Kissinger spoke of "interdependent economies," and suggested that famine "was once considered part of the normal cycle of man's existence, a local or at worst a national tragedy." Kissinger concluded, "Now our consciousness is global."[40]

A CEQ report issued in 1975 both affirmed and challenged Kissinger's ideas. While the report recognized global interconnections, it indicated, "Developed and developing countries of the world see their environmental problems differently." "Developed" countries, the CEQ believed, "were most concerned about reducing or eliminating pollution, man's threat to wildlife, the aesthetics of sprawl, and the costs of environmental degradation to man's health and quality of life." Conversely, the CEQ indicated that the "principal aspect of environmental degradation" for developing countries was "poverty, caused by lack of development." The report asserted, "The developing countries, moreover, frequently view their poverty as the result of exploitation by the affluent nations, going back to the centuries of colonial rule." As American policymakers debated a broad range of environmental initiatives on a national and global level, the CEQ report cautioned that the divergent views about the causes and consequences of environmental problems created significant problems. "The stage [is] set," the CEQ warned, "for conflict."[41]

For Kissinger and many international development experts, however, the solution to the world food crisis was not to rethink the consequences of modern agriculture, but to continue to promote technological solutions. "Existing pesticides must be made more generally available," Kissinger told a crowd. He contended that pesticides were "simple and inexpensive" and "could have a rapid and substantial impact on the world's food supply." Following the lead of Green Revolutionaries like Norman Borlaug, Kissinger firmly believed that technological solutions would banish the misery of the past. In this regard, pesticides were seen as a crucial tool in the production of food. In a 1973 speech before a workshop at the International Maize and Wheat Improvement Center, the institutional successor to the Rockefeller Foundation's Mexican Agricultural Program, Don Paarlberg, director of agricultural economics for the USDA, encapsulated the idea of agricultural development in the developing world: "Scientific and technological progress in agriculture must be accelerated, especially in the poor countries."[42] Even the CEQ claimed that "pesticides [were] part of the recommended culture."[43]

Despite the growing popularity of ecology, which only a few years earlier was thought to be the radical idea of "nature lovers," the promise of techno-

logical solutions still generated enormous enthusiasm for solving the problems of poverty, famine, and overpopulation. As ideas of interconnections
and interdependence spread through most federal agencies, and particularly
the State Department, the impulse to find technology solutions did not dis-
appear. In the realm of food production, the technological package of the
Green Revolution was thought to be the only solution to effectively feed an
expanding human population. In public health, the problem was more complicated after the dissolution of the WHO eradication program. Nevertheless,
many U.S. officials believed chemical pesticides were critical in reshaping the
world. And they were not alone.

More and more, U.S. multinational corporations, rather than philanthropic
foundations, took the lead in promoting modern agricultural technologies.
The Rockefeller Foundation continued to fund agricultural development
projects and projects related to "tropical medicine," including research on the
parasitic disease schistosomiasis. But as most Americans suffering from the
economic malaise of the 1970s were, the Rockefeller Foundation was intent
on "making less more."[44] Conversely, the opening of new international markets encouraged U.S. chemical multinationals to allocate additional resources
overseas. This shift was significant on many levels, particularly in regard to
how the logic of development changed. While the foundation promoted an
agricultural system that was largely antithetical to local agricultural practices, it did so with the best intentions to secure food for millions of people.
American multinationals, on the other hand, understood the developing
world as an emerging market.

One of the ironies of this shift, of course, is that as the United States
became more ecologically conscious, the global flow of chemical technologies increased. The results of this chemical explosion were costly, in terms
of actual dollars, ecological change, and human health. Importing nations
deploying these chemicals experienced a fundamental shift in how they produced food, since many of the chemical pesticides were used for crop production rather than for public health. From 1972 to 1985, imports of pesticides increased by 261 percent in Asia, 95 percent in Africa, and 48 percent in
South America.[45] In the Philippines, home of the International Rice Research
Institute, a Green Revolution research center sponsored by the Ford and
Rockefeller Foundations, imports of pesticides grew fivefold between 1972
and 1978.[46] According to one report, the amount of pesticides used in crop
production in Egypt jumped from 2,143 tons in 1953 to 12,550 tons in 1963
and 28,340 tons in 1978.[47] India's pesticide use increased at an annual rate of
12 percent during the 1970s and 1980s.[48] And the General Accounting Office

reported that U.S. producers exported 552 million pounds of pesticides in 1976, of which approximately 140 million pounds, or 25 percent, were unregistered.[49] Foreign governments also contributed to this incredible growth. One study on nine developing countries indicated that, on average, government subsidies covered 44 percent of the cost of pesticides and, in some cases, up to 89 percent.[50]

The human health impacts of the explosion of the global pesticide market became a "growing problem." Within the United States, over 1,300 farm workers were poisoned severely in 1975, although some estimates of the number of illnesses were as high as 100,000, since many farm workers did not seek medical attention.[51] Globally, the problem was even worse. In 1972, the WHO estimated there were approximately 500,000 cases of accidental pesticide poisoning. The New York Times reported that a small health clinic in Guatemala treated "30 to 40 a day for pesticide poisoning." One of the clinic's nurses noted, "The farmers often tell the peasants to give another reason for their sickness, but you can smell the pesticide in their clothes."[52] By 1990, the WHO reported that there were between 3.5 and 5 million cases of pesticide poisoning each year, and upward of 20,000 unintentional deaths.[53] Another study placed the figure as high as 25 million people acutely poisoned every year.[54] While it is nearly impossible to determine the full extent of poisoning during the 1970s and 1980s, there is no doubt the ecological age was one of profound technological revolution, creating enormous health risks for the people many technological enthusiasts claimed they wanted to save.

Rising malaria rates coincided with the worldwide boom of agrochemicals. During the eradication years, the number of people infected with malaria in India dropped from 100 million to 60,000 from 1952 to 1962. But by 1976, malaria affected nearly 6 million people.[55] Similar findings were made elsewhere.[56] As we have seen, the rising rates of malaria had less to do with the discontinued use of DDT than with the fallacious logic of eradication, the administrative burdens of eradication, and pesticide resistance. Indeed, the shift to organophosphates, which endangered many farm workers, was partly a response to the impact of insect resistance. Over time, the use of DDT to benefit public health and agricultural production increased the rate and risk of insect resistance. Not only did this impact crop yields, but it significantly reduced the effectiveness of DDT for malaria control programs. In Central America alone, the rate of DDT resistance was reported to be nearly 80 percent for malaria-bearing mosquitoes.[57] The interwoven histories of postwar public heath projects and agricultural development had a profound impact on the technological revolutions of the 1970s.

Recent work examining the relationship between postwar agricultural development and malaria suggests that the connections are much more dangerous and complex than previously considered. In *Maize and Grace* (2005), historian James McCann describes how African farmers during the 1990s unwittingly created environments that exacerbate the malaria problem. To generate higher yields, farmers began planting hybrid corn seeds—a Green Revolution technology—closer to villages, and in some cases they planted corn right up against the exterior walls of homes. Corn, however, is not a native crop to Africa. Its cultivation in certain parts of the continent has altered traditional growing cycles. The infusion of hybrid seed technologies only intensifies the changes to African agricultural processes.

In the Burie region of southwest Ethiopia, for example, farmers began using a late-maturing hybrid seed that was compatible with the red clay soil and the heavy summer rains. As the corn matured, it tasseled, releasing its pollen. Prior to the adoption of these hybrid seeds, tasseling occurred in July, during the height of the rainy season. The heavy rains would simply wash away the pollen.

The release of pollen from the late-maturing hybrid, however, coincided with the development stage of mosquito larvae. The larvae of the *Anopheles arabiensis* gorged on the pollen, which produced healthy adults capable of transmitting malaria. "Evidence of the timing of pollen release, the presence of increased breeding sites for larvae, and the proximity of courses of pollen to human housing," McCann indicated, "strongly suggests that the recent agroecology of maize and malaria in the Burie district has supported an increase in vector density, the range of mosquito habitat, and the bite rate, all factors leading to an outbreak of epidemic proportions." Indeed, the epidemic of 1998 saw nearly 47 percent of the population in the Burie district contract malaria. Two decades earlier, the region had "neither mosquitoes nor malaria."[58]

The entwined histories of malaria control and the Green Revolution continue to confound experts and transform human communities. They remain complex local and global issues that reinforce the unforeseen consequences of environmental change.

As Americans debated the scale and scope of pesticide regulation during the 1970s, they did so with limited knowledge of the ecological connections between public health and agricultural development. Certainly the ideology and politics of development made the links between public health and the Green Revolution readily apparent, but their ecological connections remained a mystery. Nevertheless, the pesticide debates began to scratch the surface, exposing the dangerous nexus between chemical pesticides, food production, and public health.

For many Americans who had assumed that the DDT problem had long been resolved, it must have come as a shock to find the pesticide making news once again. In December 1976, the *Washington Post* reported that the U.S. Agency for International Development (USAID) had been engaged in facilitating the transfer of chemicals banned or hazardous for domestic consumption to foreign nations.[59] In 1974, USAID purchased DDT from Montrose Chemical and shipped it to India, Ethiopia, Nepal, Indonesia, South Vietnam, and Haiti for malaria control and agricultural development projects. Not only was DDT identified as one of the chemicals, but USAID had also purchased and shipped leptophos, a chemical manufactured by Velsicol Chemical Corporation. Sold under the trade name Phosvel, leptophos had made headlines in 1976, as Velsicol workers experienced dizziness, impaired vision, and partial paralysis. These extreme nervous disorders, which gave rise to the term "Phosvel Zombies," shocked the American public when the press reported them. Leptophos had been given only a temporary registration in 1971, but was exported to various ports around the world. In one extreme case, 1,200 water buffaloes died in Egypt as a result of leptophos poisoning.[60] Passage of the Toxic Substances Control Act (TSCA) in 1976, which prohibited the manufacture or importation of nonregistered chemicals, only confounded the public's trust in the nation's regulatory system. Indeed, it seemed incredible to some that under the TSCA, U.S. chemical corporations continued to manufacture and export dangerous chemicals.

In June 1977, public furor further escalated when the Consumer Product Safety Commission (CPSC) removed tris-treated sleepwear from the market. Tris (2,3-dibromoprophyl), a flame retardant added to children's sleepwear, was found to cause cancer in lab animals. While thankful parents cautiously rejoiced at the decision, many manufacturers exported their excess inventory to avoid substantial financial losses. Tris-treated sleepwear was found in Bogotá, Caracas, and even Paris.[61] As a result of the public disclosure that the material had been exported, Congress began in 1977 a series of hearings focusing on the export and reimportation of banned substances. Triggered by the controversy over tris-treated sleepwear, an effort to find procedures to grapple with this complex issue was initiated by the federal government. The report based on the congressional hearings contended, "There is, at present, no consistent, uniform approach to U.S. export policy as it affects products banned by U.S. regulatory agencies." Likewise, concerning the reimportation of banned substances, the report found that there were no procedures in place to deal with problems, but that reimportation "should be considered a potentially serious problem and accounted for in agency decisions."[62]

USAID officials were caught in the crossfire of competing interests and regulatory confusion. Throughout the 1960s, USAID served an important function in advancing the U.S. foreign policy agenda, underwriting large parts of the Green Revolution. It also served as the funding agent supplying foreign nations with American-made DDT for eradication and other public health programs. Yet domestic environmental regulation, particularly NEPA, created a regulatory hurdle for the State Department.[63]

At issue was the requirement established in NEPA that all federal projects must undergo a comprehensive environmental review prior to the start of a project. In conducting such a study, federal agencies could better determine the short- and long-term "impacts" of a particular project.

The problem USAID faced, however, was that all of its projects were on foreign soil. How then does one determine the environmental impact of a project in a country where the risk-benefit calculus may be completely different than it is in the United States? "It raises many unanswered questions," wrote Christian A. Herter Jr., special assistant to the secretary of state for environmental affairs. "What are the consequences of polluting your neighbors? What are the damages? . . . the responsibilities? . . . the liabilities? What is the forum in which they seek redress?"[64] Some State Department officials countered that because their programs were beyond U.S. borders, those programs were not required to adhere to NEPA regulations. For others, determining the proper course of foreign projects was much more difficult.[65]

In light of these controversies, Charles Warren, CEQ chairman, asked, "What kinds of serious environmental damage can happen—and sometimes have happened—as a result of U.S. government actions outside our borders? . . . What do we stand to gain from a careful, sensitive application of the National Environmental Policy Act to the conduct of our international programs?"[66] Warren's comments only confirmed the dilemma faced by USAID, but more importantly they raised what was increasingly becoming part of the environmental policy discussion in the 1970s, the use of pesticides in the world beyond U.S. borders. In his first environmental message to Congress, President Carter addressed some of these concerns, proclaiming, "Environmental problems do not stop at national boundaries. In the past decade, we and other nations have come to recognize the urgency of international efforts to protect our common environment."[67]

To reiterate these concerns, Russell Train testified before the Committee on Foreign Relations in 1976. He asserted, "We are part of an increasingly interdependent and interrelated world. . . . There can be no thought of a retreat into isolationism." Sounding remarkably like Henry Luce thirty-five years earlier, Train understood the ecological and political connections the United States had to the rest of the world. Less clear, however, was whether the United States would take the lead in shaping global environmental policy, or how technologies of development like DDT figured into this new economic and political reality.

Nevertheless, Train's testimony emphasized the growing connection between U.S. environmental and foreign policy. The two, as Train noted, were "interdependent." But the question remained how to realize or construct a regulatory framework that considered national and international environmental issues. "International environmental cooperation," Train contended, represented "an international extension, a global dimension, of . . . domestic priorities."[68] While Train made a case for the expansion of environmental programs nationally and internationally, his comments demonstrated the belief that the world was simply a broader canvas on which to extend the "domestic priorities" of the United States. Train believed, as others had in their conceptions of the American Century, that international engagement simply expanded U.S. interests abroad. The transfer of toxic substances across international borders demonstrated the inconsistencies of U.S. environmental regulation while undermining the concept of interdependence. For Train and other like-minded environmentalists, this was a problem to be solved.

For those who claimed that DDT regulation exposed a deep-seated flaw within U.S. environmental governance, the issues were not as clear-cut as proponents of increased federal legislation believed. On the surface, "dumping" banned substances into foreign markets may have seemed dangerous and unethical, but ecological or public health factors in other countries may have warranted or demanded the use of products banned in the United States. Prior to the U.N. Vienna Conference on Science and Technology for Development in 1979, a *Washington Post* editorial offered an alternative to the "usual platitudes" and the call for "new 'institutional' mechanisms" often put forth at international conferences. Instead, the *Post* suggested that the conference find a way "to curb runaway industries and [set up] mutually acceptable controls over the export of hazardous substances." More significantly, however, the *Post* explored the controversy over banned substances in a much more nuanced way. "It's not as easy to regulate such exports as it might seem," the *Post* reminded its readers, referring to the complexity of environmental regulation in a global age. "Most health and safety regulations balance risks and benefits, and these judgments reflect values—*our* values—and generally not those of all of us at that. Imposing such decisions on others comes peril-ously close to stepping on some other country's sovereignty. Also, different environmental conditions may mitigate the nature of the hazard, acceptable alternatives may not be available or conditions may be such that a risk that looks unacceptable here seems well worth taking."[69] Political borders and regional differences, the *Post* suggested, mattered.

The universalist impulses of eradication, which supposed that a global assault against mosquitoes would produce a world without malaria, were mirrored in the beliefs of environmentalists, who also held similar universal ideals. The EDF, for one, not only acknowledged the ecological connections between diverse geographies, but also politicized these ideas to prohibit the global flow of DDT and other chemical pesticides. Moreover, the ecologi-cal ideas of wholeness and universality, perhaps best reflected in the iconic Apollo 8 image *Earthrise* (1968), became the symbol of a new environmental politics, one that imagined a "world without borders."[70]

The concept of "spaceship earth" affirmed the belief in a shared human history and an equally perilous future that transcended the politics of the nation-state. Viewed from this perspective, the earth and the limits of its resources became increasing clear, and regional differences faded amid the brilliant swirl of blues, greens, and browns. Seen from the ecological and historical perspective on the ground, however, the view looked much differ-ent. Shaped by both ecology and history, the world was not universal. Yet in the effort to find regulatory consistency on the pesticide question, "place-less and universal assertions of environmental harm," historian Christopher Sellers wrote, would deny ecological differences.[71] To put it another way, in a 1975 speech before the Nutrition Foundation describing the challenges of pesticide regulation, Train remarked, "One man's pesticide is another man's poison."[72]

Train had difficulty confronting the ethical and transborder dimensions of pesticide policy, as did many others who participated in this debate. It was not an easy problem, nor did it offer any clear-cut solutions. However, the legal and regulatory process in which pesticide policy was debated clearly shaped the tenor and tone of the debate, limiting, perhaps, a more nuanced understanding of the problem. This was certainly the case in the DDT debate. The contentiousness on all sides of the DDT issue undermined a more com-plex, and perhaps more paradoxical, view of the controversial chemical. The burdens of regulation only compounded these problems.

For the remainder of the decade, Congress continued to tinker with the international dimensions of U.S. pesticide regulations. In 1978, it mandated that foreign countries be notified in the event that hazardous materials were shipped overseas. Likewise, the FIFRA amendments of 1978 required new labels that identified the producer of the pesticide. They also required the exporter of unregistered pesticides to obtain a written acknowledgment from the foreign purchaser that stipulated full disclosure of the transaction. And lastly, labels for unregistered pesticides had to be written in two languages.[73] Of course, enforcement of these laws by multiple federal agencies would be required to ensure compliance, and recessionary budgets limited the capacity for proper oversight.

These regulatory difficulties seemed minor compared to larger structural problems that called into question the promise of the American Century. Faced with rising energy prices, deindustrialization, staggering inflation rates, and the Iran hostage crisis, many Americans thought their country's best days were behind it. The malaise and "crisis of confidence" of the Carter years marked the inglorious end of the American Century, an end marked by self-doubt, confusion, and ambiguity. According to Judith Stein, "the 1970s was the only decade other than the 1930s wherein Americans ended up poorer than they began."[74] While environmental regulation became an easy target to explain America's economic decline, the excesses of the postwar consumer republic and the forces of economic globalization, in the opinion of many, were the primary cause of the economic, political, and social upheavals of the 1970s.

The question of DDT, however, had yet to be settled. On January 15, 1981, just five days before he was to leave office, President Carter signed Executive Order 12264, which prohibited the export of banned substances from the United States. These materials included the fire retardant tris, leptophos, and DDT. Carter's order imposed additional regulatory and reporting responsibilities on the State Department, the EPA, the FDA, and the CPSC.[75] Signing the order, Carter stated, "It emphasizes to other countries . . . that they can trust goods bearing the label 'Made In U.S.A.'"[76] At a time when America's standing in the world had reached its nadir, by invoking the term "trust" and the phrase "made in the U.S.A.," Carter made a last-ditch effort to bolster the United States and its standing abroad.

The Reagan "revolution" cast the nation's malaise aside, reawakening the spirit of the American Century. Reagan's vision of the "shining city on a hill" embraced the notion of American exceptionalism, asserting that the United States was, and always had been, a force for good in the world. The technological impulses of the international development projects of the past, the Reagan revolution seemed to say, should be celebrated rather than questioned.

Reagan opposed many of the legislative advances made during the environmental decade and, upon assuming the presidency, slashed the budgets of all environmental agencies, including the EPA and the CEQ. The environmental decade had quickly and unambiguously come to a screeching halt. Reagan also reversed many of Carter's executive orders, including the prohibition on the export of banned substances. American-made DDT was, once again, free to be sold on the world market. Not everyone supported this decision. In response, Representative Michael Barnes (D-Md.) said in defiance, "The label 'Made in America' should be taken as a guarantee, not a warning."[77]

With the onset of the new decade, Montrose Chemical entered into an unfavorable global market. Increased competition from chemical interests in France and Italy, unfavorable international exchange rates, which made foreign-made DDT cheaper on the global market, and the continued presence of environmental regulations transformed the once lucrative market into a hostile battleground. As a result, in June 1982, Montrose Chemical ended production of DDT in its Torrance, California, plant. Montrose Chemical's decision to suspend DDT production signified its unwillingness to try to overcome competition from foreign manufacturers, falling prices, and increased environmental burdens.[78] In contrast, the Montrose DDT plant's closure represented the definitive triumph of environmentalists in the contentious struggle over DDT, a battle that had been twenty years in the making. Yet, while the manufacture of DDT ended in 1982, the long-term environmental impact of thirty-five years of production left substantial scars. As a result, Montrose Chemical remained embroiled in environmental litigation for the remainder of the century, often because of illegal dumping into municipal sewer systems.[79]

For all intents and purposes, DDT was no longer an American problem. The battle had been won not merely by regulatory judgments, but by the ever-shifting fluctuations of a global market. And despite the fact that DDT application was one of the most significant technologies of the American Century, little was made of its disappearance—though it never entirely "disappeared" from the environment. The international use of DDT had had a profound impact on domestic politics since the chemical first made national headlines after the Naples campaign in 1944, and DDT's absence from the United States did not much change the debate. Contemporary commentators continue to rehash old arguments about the use of DDT for malaria control work in Africa, Latin America, and Southeast Asia.

While the decision to ban the use of DDT within the United States represented a triumph of American liberalism, it also created a "backlash." And, set against a fundamental shift in America's standing abroad, concerns about the "specter of environmentalism" confirmed the belief that the conclusion of the American Century had its roots in the modern environmental movement.[80] Seen in Poage's constituents, who railed against the EPA's "dictatorial" powers, rising malaria rates around the world, and Charles Wurster's seemingly racist views of malaria sufferers, the expansion and inconsistency of American liberalism served as a counterpoint to the emergence of a powerful conservative movement that brought Ronald Reagan to the White House. And because DDT was such an important technology in defining America's overseas adventures after the Second World War, its prohibition represented a rejection of its global presence, undermining the idea of and belief in the American Century. This legacy continues to shape the politics of DDT within the United States and beyond its borders.

RETHINKING DDT IN
A GLOBAL AGE

If you were to compare the global use of DDT to the rates of malaria during the postwar period, the results would be astonishing and highly misleading. High DDT use between 1945 and 1965 produced a spectacular decrease in malaria transmission. In places like Sardinia, Sri Lanka, and India, malaria rates dropped so low that the disease appeared to have been eradicated. By the mid-1960s, however, the trend was reversed. Malaria rates rose steadily, while DDT use began to wane. Since then, DDT use has plummeted, although DDT is still in use for malaria control and as an agricultural pesticide in some parts of the world. Malaria, on the other had, has become one of the most pressing and heartbreaking public health problems of our time. The disease affects approximately 30 to 40 million people a year, resulting in the deaths of over 1.5 million people annually.

This comparison would lead to a rather unambiguous conclusion: DDT saved lives and the suspension of the spray program created a public health crisis. According to one commentator: "When residual spraying with DDT was stopped in Sri Lanka in 1964, the number of [malaria] cases increased from virtually zero to 1.5 million during a major epidemic between 1968 and 1970."[1]

Somewhat surprisingly, this observation was not made by a recent critic of the EPA, of Rachel Carson, or of the environmental movement. Instead, Rockefeller Foundation president John H. Knowles made it in the introduction to the foundation's 1975 annual report to call attention to the speed with which malaria returned to the island nation. For Sri Lankans and the international aid community it was an alarming development.

Looking at the history of DDT use and its impact on malaria, the conclusion would be rather straightforward. Eradication worked. The refusal, then, to deploy DDT because of some vague ecological threat not only unleashed a public health catastrophe of global proportions, but also revealed the narrow-mindedness of the environmental movement.

Yet like most simple distillations, this scenario hides more than it reveals. Indeed, absent from this uncomplicated depiction of DDT and malaria are the people who dreamed up the concept of eradication, the mosquitoes who resisted the deadly chemical, and the ecological processes that connected DDT to nature's web. Also lost are the politics of the Cold War, the myopia of technological modernists, and the limits of state regulation. This book attempts to undercover this lost history by adding substance and complexity to what has become a rather simple story.

And while critics continue to condemn environmentalists for their anti-DDT politics, they are, to use a rather colloquial expression, barking up the wrong tree. Environmentalists are no more responsible for the limited use of DDT for malaria prevention than are the technological enthusiasts who believed disease could be eradicated exclusively with the chemical. The causes of malaria—mosquitoes, the plasmodium parasite, poverty, ecology, and inequity—are far too complex to attribute to simply one group or to eliminate with a universal strategy. To be sure, the environmental movement shifted the debate on DDT, which made many people question the use of DDT for malaria control. However, the WHO dismantled the malaria eradication campaign well before the United States banned DDT.

Undoubtedly, humanitarian concerns underscore the fervent reaction of antienvironmentalists—epidemics should certainly cause outrage. And the more who rally to the cause, the better. However, the virulent nature of such antienvironmental criticism demands our attention. Since most of the more vociferous critics of environmental regulation have either current or past associations with the American Enterprise Institute or other "free market" organizations, their political motives are explicit. The story of DDT was one of government regulation, a story free-market advocates disliked. As we have seen, these antiregulatory beliefs began well before the creation of the EPA or the Ruckelshaus ruling. Since the Second World War, in fact, technological enthusiasts, development experts, and cold warriors have railed against critics of DDT, claiming they were, among other things, "communists," "food faddists," and "nature lovers." Since 1972, however, the criticism of environmental regulation and Rachel Carson has grown louder.

For most critics of Carson and the EPA, DDT is simply part of a larger antiregulatory politics that is only now beginning to be uncovered. Historian James Turner has documented how free-market ideology shaped environmental politics in the American West. But in a way the concepts of property rights and land use were not, the politics of DDT are about saving lives. The faces of millions of malaria sufferers reveal a public health issue that is real, profound, and undeniably complex. Malaria is not some abstract market principle; it is life and death. However, for all the antienvironmental rhetoric that continues to shape the contemporary debate over DDT, it seems that deregulation, rather than malaria control, is the desired outcome. So instead of rethinking how DDT might be useful in our own time, critics of environmentalists are more concerned with fighting past battles, suggesting that environmental regulation was the singular cause of so much harm around the world. As this book reveals, nothing could be further from the truth.

OVER THE COURSE OF working on this book, I have been asked repeatedly about my position on DDT for malaria control. Due to either my deference as a historian or my uncertainty, despite having spent years conducting research for this book, I have often given vague responses, claiming DDT offers some potential benefits but the associated risk may be too great to bear. I have never been fond of my answer, but there are no simple answers, since the history of DDT, and other pesticides for that matter, has been one of excess, rather than caution. DDT's history has been one of unbridled enthusiasm for a technology many thought would rid the world of disease in the span of a decade. And it has been a history shaped by universal claims about disease, famine, and environments.

DDT, as an indoor residual spray, could lower global malaria rates without causing much ecological damage. In conjunction with other forms of mosquito and disease control measures—bednets, drug therapies, drainage, cleaner water, better housing—the use of DDT could further reduce malaria transmission. And if these control measures were designed with local conditions in mind, if public health officials recognized that the risk-benefit analysis differs from place to place, then perhaps the use of DDT would be prudent. In some areas, DDT might not work at all due to continued mosquito resistance or because other local conditions might prevent the chemical from being an effective means of mosquito control. Nor should chemical control be the first line of defense against mosquitoes. Strategic thinking is necessary to prevent the unnecessary use of the pesticide. And to suggest that DDT, used by itself or deployed within the framework of a large-scale, institutionalized malaria campaign, would yield long-term success would be extremely problematic.[2]

If history is any guide, then we need to think seriously about the potential of DDT as a public health tool. Acknowledging the material realities of nature, while assessing the risks and benefits associated with the use of any chemical, DDT included, could lead to a more reasonable rethinking of the relationship between insects and humans. We also must be wary, perhaps, of the slippery slope that might occur if we fail to rethink our basic assumptions about the transformative possibilities of technology. The zealotry in which technological enthusiasts unleashed DDT without much forethought or hindsight remain a powerful force behind recent calls to reintroduce the chemical.

One only has to consider a *Wall Street Journal* opinion piece from 2007 as a cautionary tale. In an editorial on the "uses of DDT," Henry I. Miller, a fellow at the Hoover Institution and a former official at the National Institutes of Health and the FDA, claimed, "Thanks to flawed, politically correct federal regulatory policy . . . the available tools are limited and largely ineffective." Miller argued for the use of DDT, not for malaria control in developing countries, but to control an outbreak of West Nile virus within the United States.[3] The recent bedbug infestation in the United States also has led many people, including the shock-jock Howard Stern, to call for DDT spraying. Stern announced, "They gotta bring back DDT. Stop being a bunch of pussies. Seriously! Screw the bald eagle and the fragile ecosystem, bedbugs have turned up in Howard's limo!"[4]

No stranger to hyperbole, Stern nevertheless was not alone in his assessment. In fact, shortly after his initial rant, he returned to the subject armed with evidence in the form of a recent article written by Paul Driessen, a senior fellow at the conservative Atlas Economic Research Center and the author of the book *Eco-Imperialism: Green Power, Black Death*. An outspoken critic of the modern environmental movement, Driessen lashed out at what he saw as the continued hypocrisy of DDT politics. His article, "New York's bedbugs vs. Africa's malaria crisis," invoked a familiar theme in his writings, namely that American liberals are more concerned with their material comforts than with saving millions of Africans from the plague of malaria. "It is hellish for people who must live with bedbugs, and can't afford professional eradication like what Hilton Hotels or Mayor Bloomberg might hire," Driessen wrote. "But imagine what it's like for two billion people who live 24/7/365 with insects that definitely are responsible for disease: malarial mosquitoes." Of course, for Driessen, environmentalists are ultimately responsible for both problems: "These zealots," he protested, "are making decisions that determine the quality of life for millions of Americans, especially poor families—and life itself for billions of malaria-threatened people worldwide."[5]

These observations point out that DDT is not some artifact relegated to the dustbin of history. It is a chemical that is very much in the present and quite likely in our future. Is it possible, therefore, to think that DDT warrants prohibition for its ecological hazards while the benefits of DDT for malaria control may justify its use? Can we recognize that even though ecology points out that we're all connected—humans, nature, and nations—that there are vast differences between and within the environments and countries where malaria is endemic and those where it is not? In the past, universal assertions of harm, defined either for human health or for the environment, concealed how local conditions created distinct problems. And it is in the local environments that the battle of malaria ultimately may be won. It may be that big problems like malaria may not need big solutions, but rather solutions rooted in local contexts that take into consideration history, culture, politics, and nature. International aid organizations certainly have a role to play, particularly in raising and allocating funds. But given the size and scope of the malaria problem, small solutions may yield big results.[6]

The author and journalist Sonia Shah has recently suggested that malaria may never be extinguished, so humans may have to learn to live with the disease. "While eradication requires abrupt interventions to break the cycle of transmission, sustained just long enough for the parasite to die out," Shah writes, "learning to live with malaria means working to permanently sever the connections between mosquitoes and humans."[7] However noble the idea of eradication may seem, the lessons of the past tell us that short-term goals fail to generate long-term improvements in health. Certainly there are exceptions to this history, but they are few and far between. Eradication, therefore, should not be the model the public health community aspires to. The current global crisis requires long-term commitments on the part of the international community to help local efforts to reduce the rates of infection and alleviate the structural causes of disease. DDT may be part of the solution, but whether it is largely depends on local conditions. To consider how the pesticide might be used for eradication would merely replicate past failures.

But there is more to the problem than simply arguing for or against the use of DDT for antimosquito work. The scope of the pesticide problem, not simply DDT, has exploded over the course of the twentieth century. And while many debate the merits or consequences of prohibiting "persistent organic pesticides," the global use of organophosphates continues to grow. In 2001, the global consumption of pesticides reached nearly 5 billion pounds, with the United States consuming nearly 24 percent of this total.[8] According to one report, within the United States, "acute pesticide poisoning in the agricultural industry continues to be an important problem," suggesting the need "for heightened efforts to better protect farm workers from pesticide exposure."[9] Worldwide, the problem is worse. In some "developing" countries, the World Health Organization reported, agricultural workers are poisoned at a rate as high as 18.2 per 100,000.[10] And, of course, the 1984 disaster at the Union Carbide factory in Bhopal, India, which killed over 15,000 people and injured nearly 500,000, was a horrific moment in a larger historical process involving the history of DDT and the American Century.

The belief in technology's ability to radically transform agricultural systems throughout the world has increased crop yields to historic levels, as Norman Borlaug pointed out repeatedly before his death in 2009. However, basic grains are more likely to be used to feed livestock or create industrial foods like high-fructose corn syrup than to appear in the consumer market in their natural form. At the same time, the transfer of Green Revolution technologies, particularly chemical pesticides, increased the health risks to those who grew crops. Not surprisingly, those advocating for the use of DDT for malaria have had little to say about the risks to agricultural workers around the world. Since the principles of the free market stimulate the transborder flow of pesticides, many critics of DDT regulation simply overlooked the deadly consequences of the industrial agriculture system, even though it has become the predominate method of food production.

Given the contested history of the Green Revolution and the global impact of the U.S. and European pesticide industry, that some "developing" countries rejected the latest technological innovation, "genetically modified seeds," is not surprising. As the Monsantos of the world attempt to redefine the rules of food production, corporate interests within and beyond U.S. borders continue to challenge nations to grapple with a myriad of issues related to food, health, and the environment. As we have seen, these issues are not new, but the evolution of agricultural technologies, including the trademarking of seeds, only adds to the complexity of an already complex system of food production.

What is certain, however, is that the politics of DDT remain as contentious as they were when DDT was first introduced to the marketplace in 1945. And because the chemical redefined the relationship between humans and nature, the technology was infused with a meaning that became more important than its ability to kill insects. Indeed, as an integral part of the American Century, DDT fulfilled the hopes and aspirations of a generation of technological enthusiasts who embraced the transformative possibilities of the chemical. It became a symbol of hope and America's beneficent empire amid the tensions of the Cold War. The ecological impacts of the pesticide, however, exposed the flawed logic of technology. While it killed insects, DDT also despoiled the environment and created long-term health impacts, many of which we have yet to fully understand.[11]

In an effort to shed light on DDT's contested and multifaceted history, I have emphasized how ideas about ecology and technology had political implications beyond the laboratory. Science, throughout the American Century, became increasingly popular and political. I have also argued that context matters. Because DDT was such a powerful instrument of the Cold War, that power affects how we understand the chemical's history. Therefore, to better understand the full scope of DDT's history from an American perspective, we need to look beyond national borders, to draw on both the international and national dimensions of the chemical's history. It is this context that also requires us to reflect on the contentiousness that shapes the contemporary debate.

Given that DDT was a critical aspect of the American Century, what does it mean when many begin to suggest that the American Century has come to an end? Was the regulation of DDT symbolic of America's international decline? Was Reagan's reversal of many of the policy initiatives from the "environmental decade" predicated simply on his opposition to regulation, or on his commitment as well to reawakening the American Century after the tumult of the 1960s and 1970s? Was Montrose's inability to compete in a global marketplace indicative of a large shift in capital, which transformed the industrial strength of the U.S. postwar economy into a globalized, service-oriented marketplace? It is not surprising that India and China are the primary producers of DDT today, with India being the largest manufacturer of the chemical.[12] The intertwined histories of the American Century and DDT shed light on the contentious debate about DDT's use within the United States. As the United States grapples with what political scientist David Mason calls "The End of the American Century," these issues have become more pronounced and problematic.[13]

In 2008, however, the WHO changed its policy on DDT. The organization "reaffirmed the importance of indoor residual spraying (IRS) as one of the primary interventions for reducing or interrupting malaria transmission in countries where this intervention is appropriate, including in situations of high malaria transmission." The use of DDT would be limited to "indoor residual spraying only."[14] While spray programs remain somewhat limited, a new chapter of DDT's history is clearly being written. What is unclear, however, is whether the lessons of the past will be learned and what role, if any, the United States will play in DDT's twenty-first-century history.

Notes

Introduction

1. Gen. Gnassingbe Eyadema, president of the Togolese Republic, http://www.rollbackmalaria.org/docs/AMD/quotes.htm.

2. Roll Back Malaria, "World Malaria Day 2010: Africa Update: Progress & Impact Series," 2 (April 2010): 5. Report can be found at http://www.unicef.org/media/files/rbm-reportII-en.pdf.

3. Roberts and Tren, *The Excellent Powder*.

4. "Fish Killed by DDT in Mosquito Tests," *New York Times*, August 9, 1945, 23.

5. United Nations, Stockholm Convention on Persistent Organic Pollutants, at http://chm.pops.int/default.aspx.

6. Michelle Malkin and Michael Fumento, "Rachel's Folly: The End of Chlorine," Competitive Enterprise Institute, March 1996, 14.

7. Jenny Zhang, "Malaria Eradication in Sri Lanka: A Critical Analysis of the First WHO Strategy for Global Malaria Eradication," February 18, 2010, http://historyofmalaria.com/2010/02/malaria-eradication-in-sri-lanka-a-critical-analysis-of-the-first-who-strategy-for-global-malaria-eradication.

8. Roger Bate, testimony, Senate Committee on Homeland Security and Government Affairs, Subcommittee on Federal Financial Management, Government Information, and International Security, May 12, 2005. As part of his presentation, Bate submitted a working paper titled "The Blind Hydra: USAID Policy Fails to Control Malaria," coauthored with Ben Schwab, to the subcommittee records. This paper is available at http://www.aei.org/paper/22355.

9. Driessen, *Eco-Imperialism*, 28. Many of J. Gordon Edwards's views, and his claim regarding DDT, can be found at http://www.junkscience.com/ddtfaq.html.

10. See http://www.rachelwaswrong.org.

11. On environmentalists' supposed indifference to DDT's lifesaving ability, see Walter Williams, *"Silent Spring*: RIP 2004," *Capitalism Magazine*, July 7, 2004, http://capmag.com. Williams's argument quotes Charles Wurster, one of the founding members of the Environmental Defense Fund, who portrayed environmentalists in a negative light. This quote has often been misread and misused. I explore this issue in Chapter 6. For original testimony, see Committee on Agriculture, House of Representatives, February and March 1971, CIS NO 71-H161-3, 266–68. Crichton, *State of Fear*, 487; Letter from Senator Tom Coburn to Representative Jason Altmire, June 5, 2007, available at http://coburn.senate.gov/public/index.cfm/2007/6/dr-coburn-stands-for-science-rachel-carson-and-the-death-of-millions; and "About the Film," *Three Billion*

and *Counting*, http://www.3billionandcounting.com/about.php. For a brief yet excellent summary of the vilification of Rachel Carson, see Kirsten Weir, "Rachel Carson's Birthday Bashing," *Salon*, June 29, 2007, http://www.salon.com/news/feature/2007/06/29/rachel_carson.

12. Henry R. Luce, "The American Century," *Life*, February 17, 1941, 61–65.

13. Historians of technology have explored these issues elsewhere: Hughes, *Human-Built World*; Misa, Brey, and Feenberg, *Modernity and Technology*; Adas, *Dominance by Design*; and especially Johan Schot, "The Contested Rise of a Modernist Technology Politics," in Misa, Brey, and Feenberg, *Modernity and Technology*, 257–78.

14. I have written about ecological consciousness elsewhere; see Kinkela, "The Ecological Landscapes of Jane Jacobs and Rachel Carson." Also see Berman, *All That Is Solid Melts into Air*, 315.

15. Michael Bess has written probably the best book on the subject of technological and ecological modernity; see Bess, *Light-Green Society*.

16. Transcript of an oral history interview with Edward Knipling conducted by Edmund P. Russell, September 10, 1992 (part 1 of 2), p. 13, box 1, folder 24, 1992, Edward Fred Knipling Papers, Special Collections of the National Agricultural Library, Beltsville, Md.

17. Zachary, *Endless Frontier*, 1. For an insightful analysis of the connection between health and national security, see King, "The Influence of Anxiety"; King, "Security, Disease, Commerce: Ideologies of Post-Colonial Global Health"; Price-Smith, *The Health of Nations*; and Price-Smith, *Contagion and Chaos*.

18. In *The Influence of the Carnegie, Ford, and Rockefeller Foundations on American Foreign Policy*, Edward H. Berman argues, "The Carnegie, Ford, and Rockefeller foundations have consistently supported the major aims of United States foreign policy, while simultaneously helping to construct an intellectual framework supportive of that policy's major tenets" (5). Also see Cooper and Packard, *International Development and the Social Sciences*, 1; Mitchell, *Rule of Experts*; and Scott, *Seeing Like a State*.

19. For work on internationalizing U.S. history, see Bender, *A Nation Among Nations*; Rauchway, *Blessed Among Nations*; and a special issue of the *Journal of American History* 86, no. 3 (December 1999), edited by David Thelen and called "The Nation and Beyond." For a historiography on works internationalizing U.S. history, see Sexton, "The Global View of the United States."

20. Worster, "World Without Borders."

21. For a good analysis and overview of these boundary issues, see Taylor, "Boundary Terminology."

22. For a recent examination of the transnational turn, see Briggs, McCormick, and Way, "Transnationalism: A Category of Analysis," and Seigel, "Beyond Compare: Comparative Method after the Transnational Turn."

23. Charles Bright and Michael Geyer, "Where in the World Is America?" in Bender, *Rethinking American History in a Global Age*, 68.

24. McAlister, *Epic Encounters*, 1.

Chapter 1

1. Congressional Research Service Report for Congress, "Life Expectancy in the United States," August 16, 2006, http://aging.senate.gov/crs/aging1.pdf.

2. Grey, *New Deal Medicine*. The Civilian Conservation Corp built much of its program on the health of its members; see Maher, *Nature's New Deal*, esp. 115–51. Also see Smith, *Sick and Tired of Being Sick and Tired*, esp. 58–85.

3. Leavitt and Numbers, *Sickness and Health in America*. For more on the history of health in the United States, see Rosenberg, *The Cholera Years*; Humphreys, *Yellow Fever and the South*; and Grob, *The Deadly Truth*.

4. *Malaria Control in War Areas, 1942–43*, U.S. Public Health Service, Atlanta, Georgia.

5. Russell, introduction to *Preventative Medicine in World War II*, 2.

6. Zinsser, *Rats, Lice and History*, 153.

7. Cushing, *History of Entomology in World War II*, 1–10; Zinsser, *Rats, Lice and History*, 150–65; J. R. McNeill, *Mosquito Empires*, 209–34; Robert Rice, Reporter at Large, "DDT," *New Yorker*, July 17, 1954.

8. Thomas Parran, "Public Health Implications of Tropical and Imported Diseases: Strategy against the Global Spread of Disease," *American Journal of Public Health and the Nation's Health* 34, no. 1 (January 1944): 1–6.

9. Robert A. Moore, "Medical Internationalism," *Scientific Monthly* 61, no. 3 (September 1945): 173–80.

10. Advertisements in *Soap and Sanitary Chemicals*, January and February 1944.

11. Tsutsui, "Landscapes in the Dark Valley," 297–311.

12. Pat Jarrett, "The Story of Pyrethrum," *Agricultural Chemicals* vol. 1 (December 1946), 27–30, 49.

13. Quoted from Russell, *War and Nature*, 127. Russell's book is perhaps the most comprehensive study of the military control of insect pests. As he accurately demonstrates, the language of eradication and annihilation served a dual purpose that merged military and public health work during the Second World War. I argue that health rather than annihilation is an important trope to examine in this period.

14. See Perkins, *Insects, Experts, and the Insecticide Crisis*, 3–15, 98–103; Perry Adkisson and James Tumlinson, "Edward F. Knipling, 1909–2000," in *Biographical Memoirs*, vol. 83 (Washington, D.C.: National Academies Press, 2003), 1–15; Edward F. Knipling, "Autobiographical Sketch," *McGraw-Hill Men of Science* 2 (1968): 296–98; Biographical Sketch (Extended), Special Collections at the National Agricultural Library: Edward Fred Knipling Papers: Screwworm Eradication Program Records, http://www.nal.usda.gov/speccoll/collectionsguide/knipling/210ExtBio.pdf. Also see Edward B. Knipling, "The Life and Vision of Edward F. Knipling Concerning the Eradication of the Screwworm," http://www.ars.usda.gov/aboutus/docs.htm.

15. Executive Order 9285, Establishing the United States of America Typhus Commission, December 24, 1942. Also see Oshinsky, *Polio*, 101–3, and Slater, *War and Disease*.

16. From Geigy press conference, *Soap and Sanitary Chemicals* 20 (May 1944). Also

see Victor Froelicher, "The Story of DDT," *Soap and Sanitary Chemicals* 20 (July 1944): 114–19, 145.

17. Edward F. Knipling, "The Development and Use of DDT for the Control of Mosquitoes," *Journal of the National Malaria Society* 4, no. 2 (June 1945): 78. Also see Edward F. Knipling, "Insect Control Investigations of the Orlando, Fla., Laboratory During World War II," *Annual Report of the Board of Regents of the Smithson Institution*, June 30, 1948, 331–48.

18. See "U.S.D.A. Set-Up For Testing Army Louse Powder," *Soap and Sanitary Chemicals* 19 (1943). Also see "The Army Tests its New Insecticide," *Soap and Sanitary Chemicals* 20 (March 1944).

19. Lukas Straumann and Daniel Wildmann, *Schweizer Chemieunternehmen im "Dritten Reich" (Swiss Chemical Enterprises in the "Third Reich")* (Zurich: Chronos, 2001). Also see "DDT Still has its Place," April 17, 2003, a radio program with Lukas Straumann, available at www.swissinfo.org; Raymond G. Stokes, review of *Schweizer Chemieunternehmen im "Dritten Reich"* in *American Historical Review* 109, no. 1 (February 2004): 272; and Eckart and Vondra, "Malaria and World War II." While there is little proof that the Allies engaged in biological warfare, Nazi leaders, who themselves deployed chemical and biological weapons, believed in the real possibility of such an attack.

20. Stapleton, "The Short-Lived Miracle of DDT."

21. R. C. Roark, "Early DDT Investigations in Foreign Countries," *Agricultural Chemicals* (July 1946), 21–23, 57–61.

22. Weindling, *Epidemics and Genocide in Eastern Europe, 1890–1945*, 373.

23. Joy, "Malaria in American Troops in the South and Southwest Pacific in World War II," 193.

24. Webb, *Humanity's Burden*, 3–12.

25. Humphreys, *Malaria*, 55–60.

26. Zinsser, *Rats, Lice and History*; Joseph M. Conlon, "The Historical Impact of Epidemic Typhus," unpublished manuscript, available at http://entomology.montana.edu/historybug/typhus-conlon.pdf; and Baumslag, *Murderous Medicine*, 9–21.

27. Edward F. Knipling, "Insect Control Investigations of the Orlando, Fla., Laboratory During World War II," *Annual Report of the Board of Regents of the Smithson Institution*, June 30, 1948, 331–48.

28. Edward F. Knipling, "The Development and Use of DDT for the Control of Mosquitoes," *Journal of the National Malaria Society* 4, no. 2 (June 1945): 77. Also see R. C. Bushland et al., "Development of a Powder Treatment for the Control of Lice Attacking Man," *Journal of Parasitology* 30, no. 6 (December 1944): 377–87.

29. R. C. Bushland, L. C. McAlister Jr., G. W. Eddy, and Howard A. Jones, "DDT for the Control of Human Lice," *Journal of Economic Entomology* 73, no. 1 (February 1944): 126–27. Also see Arthur W. Lindquist, A. H. Madden, H. G. Wilson, and Howard A. Jones, "The Effectiveness of DDT as a Residual Spray against Houseflies," *Journal of Economic Entomology* 73, no. 1 (February 1944): 132–34.

30. Edward F. Knipling and W. E. Dove, "Recent Investigations of Insecticides and

Repellents for the Armed Forces," *Journal of Economic Entomology* 37, no. 4 (1944): 477–80; Edward F. Knipling, "DDT to Control Insects Affecting Man," *Journal of Economic Entomology* 39, no. 3 (1946): 360–66; Edward F. Knipling, "DDT Insecticides Developed for Use By the Armed Forces," *Journal of Economic Entomology* 38, no. 2 (1945): 205; A. H. Madden, A. W. Lindquist, and Edward F. Knipling, "DDT and Other Insecticides as Residual-type Treatments to Kill Bedbugs," *Journal of Economic Entomology* 38, no. 2 (1945): 265–71; A. H. Madden, A. W. Lindquist, and Edward F. Knipling, "DDT as a Residual Spray for the Control of Bedbugs," *Journal of Economic Entomology* 37, no. 1 (1944): 127–28; A. W. Lindquist, A. H. Madden, H. G. Wilson, and Edward F. Knipling, "DDT as a Residual-type Treatment for Control of House Flies," *Journal of Economic Entomology* 38, no. 2 (1945): 257–61; and A. W. Lindquist, A. H. Madden, and Edward F. Knipling, "DDT as a Treatment for Fleas on Dogs," *Journal of Economic Entomology* 37, no. 1 (1944): 138.

31. United States Department of Agriculture, Bureau of Entomology and Plant Quarantine, Interim Report 0-100, "Insecticides and Insect Repellents Developed for the Armed Forces at the Orlando, Fla. Laboratory," Sept/Oct 1945, folder 1989, box 292, series 200, RG 2, Rockefeller Foundation Archives, Rockefeller Archive Center, Sleepy Hollow, N.Y. (hereafter RAC).

32. *Rockefeller Foundation Annual Report*, 1944, 64, 87.

33. Stapleton, "A Lost Chapter in the Early History of DDT," 522.

34. Percy N. Annand to John A. Ferrell, April 5, 1944, folder 1813, box 362, series 200, RG 2, RAC.

35. William A. Davis and Charles M. Wheeler, "The Use of Insecticides on Men Artificially Infested With Body Lice," *American Journal of Hygiene* 39 and 40 (1944): 163–76.

36. Pre-DDT louse-dusting research was conducted by the Rockefeller Foundation's IHD. See William A. Davis, Felipe Malo Juvera, and Pilar Hernandez Lira, "Studies on Louse Control in a Civilian Population," *American Journal of Epidemiology* 39, no. 2 (1944): 177–88.

37. George Payne to Hugh Smith, October 22, 1945, folder 149, box 18, series 313I, RG 1.1, RAC.

38. My use of the term "inscribe" is influenced by Latour, *Pandora's Hope*, 174–215, and Lenoir, *Inscribing Science*, esp. 1–19.

39. For a small sample of recent literature, see Humphreys, *Malaria*; Anderson, *Colonial Pathologies*; Molina, *Fit to Be Citizens?* esp. 179–88; and Shah, *Contagious Divides*.

40. Humphreys, *Yellow Fever and the South*, 15.

41. See the short biography "Fred L. Soper" from the National Library of Medicine, http://profiles.nlm.nih.gov/ps/retrieve/Narrative/VV/p-nid/76.

42. Fred Soper, "Paris Green in the Eradication of *Anopheles Gambiae*: Brazil, 1940; Egypt, 1945," *Mosquito News* 26, no. 4 (December 1966).

43. Packard and Gadelha, "A Land Filled with Mosquitoes."

44. Smith, "Mustard Gas and American Race-Based Human Experimentation in

World War II." For more on the relationship between science, human experimentation, and chemical weapons, see Goliszek, *In the Name of Science*; Levy and Sidel, *War and Public Health*.

45. Raymond C. Shannon to Wilbur A. Sawyer, July 10, 1943, RG2, series 200, box 245, folder 1695, RAC.

46. Stapleton, "The Dawn of DDT and Its Experimental Use by the Rockefeller Foundation in Mexico, 1943–1952."

47. Dr. Ferrell to M. F. Boyd, July 28, 1942, RG 1.1, series 323I, box 18, folder 148, RAC.

48. E. F. Knipling to George Strode, July 20, 1943, RG 2, series 200, box 245, folder 1965, RAC; Howard Hutter to George Strode, February 11, 1994, RG 2, series 200, box 263, folder 1813, RAC; also see André Luiz Vieira de Campos, "Combatendo Nazistas e Mosquitos: Militares Norte-Americanos no Nordeste Brasileiro, 1941–1945," *História, Ciências, Saúde-Manguinhos* 5, no. 3 (November 1998–February 1999): 603–20.

49. *Ventures in World Health*, 240.

50. See the influential work of Mitchell, *Rule of Experts*, 27. Mitchell's first chapter, "Can the Mosquito Speak?" examines the interactions between mosquitoes and the mapping of the Nile River Valley by scientific experts.

51. Regarding the shipment of DDT to Egypt, see George Strode to Paul Soper, November 13, 1944, folder 1813, box 263, series 200, RG 2, RAC.

52. Memorandum from Fred L. Soper to William S. Stone, October 29, 1943, box 65, Fred L. Soper Papers, National Library of Medicine, Bethesda, Maryland (hereafter NLM).

53. Fred L. Soper to Brigadier General Leon A. Fox, USATC, September 2, 1943, Fred L. Soper Papers, NLM.

54. Edward F. Knipling, "The Development and Use of DDT for the Control of Mosquitoes," *Journal of the National Malaria Society* 4, no. 2 (June 1945): 77–92.

55. George Strode to Paul F. Russell, March 22, 1943, folder 1695, box 245, series 2, RG 2, RAC.

56. USDA, Bureau of Entomology and Plant Quarantine, Interim Report 0-100, "Insecticides and Insect Repellents Developed for the Armed Forces at the Orlando, Fla., laboratory," September/October 1945, p. 75, folder 1989, box 292, series 200, RG 2, RAC.

57. Herbert L. Matthews, "Naples Goes Mad with Joy as Grim Allied Push Ends," *New York Times*, October 2, 1943.

58. Cushing, *History of Entomology in World War II*, 59.

59. Stapleton, "The Short Lived Miracle of DDT." Also see Charles M. Wheeler, "Control of Typhus in Italy 1943–1944 by Use of DDT," *American Journal of Public Health* 36 (February 1946): 119–29. Fred L. Soper, W. A. Davis, F. S. Markham, and L. A. Riehl, "Typhus Fever in Italy, 1943–45, and Its Control with Louse Powder," *American Journal of Hygiene* 45, no. 3 (May 1947): 305–34.

60. Frederick C. Painton, "The Second Battle of Naples—Against Lice," *Reader's Digest*, June 1944, 21–22.

61. "The Conquest of Typhus," *New York Times*, June 4, 1944.

62. Fred L. Soper to Wilbur Sawyer, February 7, 1944, Fred L. Soper Papers, NLM.

63. Joy, "Malaria in American Troops in the South and Southwest Pacific in World War II."

64. See Cushing, *History of Entomology in World War II*, 41–42.

65. Russell, introduction to *Preventative Medicine in World War II*.

66. For Russell's less than enthusiastic support for DDT, see Humphreys, *Malaria*, 149–57.

67. Russell, introduction to *Preventative Medicine in World War II*, 5–6.

68. See *Preventative Medicine in World War II*, vol. 9 (Washington, D.C.: Office of the Surgeon General), 55–84.

69. For more on malaria efforts in the Pacific, see Major Thomas A. Hart, "The Army's War Against Malaria," *Scientific Monthly* 62, no. 5 (May 1946): 421–22.

70. Ibid. Also see Hays, "The United States Army and Malaria Control in World War II."

71. Bennett, "Malaria, Medicine, and Melanesians."

72. "Malaria Incidence in Army is Cut 75%," *New York Times*, November 4, 1944.

73. "The Troops: Manpower Balance," *New York Times*, October 10, 1945.

74. James I. Terwilliger, "Medicine and Science Triumph over Malaria: Defeat of Anopheles," World War II Photos and Documents—14th Malaria Control Division—Dobadura, New Guinea; Owi Island, Netherlands East Indies; Leyte and Manila, Philippines—1942–1945 / 1944-06-Defeat-of-Anopheles-0001, http://linkage.cpmc.columbia.edu/WWII/index.html.

75. The report from 1944–45 from the Office of Malaria Control in War Areas published a timeline of the Second World War, including key moments of the agency's malaria campaigns, against the backdrop of a mushroom cloud. See *Malaria Control in War Areas, 1942–43*, U.S. Public Health Service, Atlanta, Ga.

76. Jane Stafford, "Insect War May Backfire," *Science News Letter*, August 5, 1944, 91.

77. James S. Simmons, "How Magic is DDT?" *Saturday Evening Post*, January 6, 1945, 85.

78. From Andrew J. Warren officer's diary regarding visit to noted malariologists, M. A. Barber, folder 1695, box 245, series 200, RG 2, RAC.

79. "Advice on DDT Toxicity Hazard," *Soap and Sanitary Chemicals* 21 (November 1945): 129. Also see "Health Hazards of DDT," *Soap and Sanitary Chemicals* 21 (August 1945): 127.

80. "DDT Warning," *Time*, August 7, 1944, 66.

81. "DDT Dangers Discussed in Expert's Talk," *Christian Science Monitor*, October 20, 1945; "DDT Spray Called Injurious to Birds," *New York Times*, October 23, 1945; John H. Baker, "DDT, Caution!" *Audubon*, September 1945, 311–12; B. Jones, "DDT: Handle with Care," *Better Homes and Gardens*, November 1945, 10; J. Feiner et al., "DDT will get 'em, but . . . ," *Ladies Home Journal*, June 1946, 176–77.

82. Lawrence Griswold, "DDT Could Upset the Economy of Nature," *Christian Science Monitor*, August 16, 1945.

83. "Doubts DDT Caused Bay Fish Deaths," *Washington Post*, August 23, 1945.

84. Arthur Hays Sulzberger to Raymond Fosdick, January 12, 1945, folder 1989, box 292, series 200, RG 2, RAC.

85. This was the general argument of Russell's insightful article, "The Strange Career of DDT," 776–81.

86. Henry Nelson, speech reprinted in *Soap and Sanitary Chemicals* 20 (June 1944): 110.

87. John Rodda, "Insecticide Outlook," *Soap and Sanitary Chemicals* 21 (August 1945): 112.

88. Baxter, *Scientists Against Time*, 369. In 1943, 153,000 pounds of DDT were produced. Ten million pounds were produced in 1944.

89. Advertisements, *Soap and Sanitary Chemicals* 21 (September 1945): 118, and (October 1945): 100.

90. Baxter, *Scientists Against Time*, 376.

91. Statement by Brigadier General James S. Simmons before the Senate Subcommittee on Wartime Health and Education, December 14, 1944, folder 732, box 61, series 200, RG 1.1, RAC.

92. "DDT Outlook . . . ," *Soap and Sanitary Chemicals* 20 (June 1944): 127–29, 147. A reprint of Victor Froelicher's comments can be found in "The Story of DDT," *Soap and Sanitary Chemicals* 20 (July 1944):, 115–19, 145.

93. George Strode to Paul N. Annand, June 12, 1948, RG 2, series 200, box, 413, folder 2782, RAC.

94. "Golden Age of Entomology," *Journal of Economic Entomology* 37, no. 1 (February 1944): 160.

95. Vannevar Bush, "As We May Think," *The Atlantic*, July 1945.

Chapter 2

1. Introduction, *Fortune*, July 1934, 45. Also see Augspurger, "Henry Luce, *Fortune*, and the Attraction of Italian Fascism."

2. Brinkley, *The Publisher*, 352–90.

3. Hanhimäki, "Global Visions and Parochial Politics." Although Hanhimäki explores the notion of parochialism in the 1960s, the term captures the essence of Luce's critique of American public sentiment toward international engagement.

4. Luce, "The American Century."

5. Snowden, *The Conquest of Malaria*, 142–80.

6. "The State: Fascist and Total," *Fortune*, July 1934, 46.

7. Caprotti and Kaïka. "Producing the Ideal Fascist Landscape"; Federico Caprotti, "Malaria and Technological Networks," 145–55; Snowden, The *Conquest of Malaria*, 184–85.

8. Snowden, *The Conquest of Malaria*, 185.

9. Snowden argued that this was a form of biological warfare. See *The Conquest of*

Malaria, 184–97. Geissler and Guillemin offer a different interpretation; see "German Flooding of the Pontine Marshes in World War II," 2–23.

10. Joseph J. Thorndike Jr., "Food: Postwar World Needs Twice as Much Food to Lift Universal Blight of Malnutrition," *Life*, October 4, 1943, 96–97.

11. Kozol, *Life's America*, 5.

12. I employ Deborah Poole's term to call attention to the transborder consumption of visual texts and the centrality of race and racial construction within this international economy. See Poole, *Vision, Race, and Modernity*, 8–25. For more on images and racial construction, see Wexler, *Tender Violence*, esp. 52–93.

13. Brandt was part of a legion of photographers who traveled throughout the world to document the war. On June 6, 1944, Brandt embarked on a remarkable, if not historic, assignment. Traveling with the Allied invasion force across the English Channel, Brandt would capture one of the first images of the Normandy Invasion. Upon returning to London, Brandt's photograph, according to *Life*'s picture editor, John G. Morris, was "not a terribly exciting one." It was a picture "of a momentarily unopposed landing on the French coast, shot from the bow of his landing craft." While Brandt's image certainly did not capture the chaos and violence of the Normandy Invasion in the same way that Robert Capra's images would—a set of images that became some of the most powerful in the history of photojournalism—he was part of a historic event that changed the nature of the war (Morris, *Get the Picture*, 7). Also see Moeller, *Shooting War*, 196–99; Paul Healy, "Cameramen Under Fire," part 2, *Popular Mechanics*, November 1944, 65–69, 152–54; and Ernie Pyle, "Ernie Helps Pose Pictures in Liberated Normandy," *St. Petersburg Times*, June 29, 1944.

14. Allen Raymond, "Now We Can Lick Typhus," *Saturday Evening Post*, April 22, 1944, 14–15, 70.

15. "DDT," *Time*, June 12, 1944, 66, 68; "Super-Delouser," *Newsweek*, June 12, 1944, 96. Plasma and penicillin were the other two great inventions.

16. After a thorough search, I found that the archives at Time-Life did not contain any additional material from Brandt's work in Naples, so determining the images that were not published proved difficult. Two additional sources provided some assistance. The Corbis archive holds a number of Brandt's photographs of the Naples typhus campaign. The Houston Metropolitan Research Center at the Houston Public Library houses the Brandt archive. Timothy Ronk and Joel Draut at the library were most helpful.

17. Ideas about filth, disease, and excluded groups were not new. For insightful analysis of these issues, see Cockayne, *Hubbub*; Burnstein, *Next to Godliness*; and Tomes, *The Gospel of Germs*.

18. See Marco Mariano, "The U.S. Discovers Europe: *Life* Magazine and the Invention of the 'Atlantic Community' in the 1940s," in Vaudagna, *The Place of Europe in American History*, 161–86. Also see Marco Mariano, "Is Italy an 'Atlantic Country'?" (working paper, Italian Academy for Advanced Study in America, Columbia University, April 2004.), www.italianacademy.columbia.edu/publications/

working_papers/2003_2004/paper_fa03_Mariano.pdf; and Mariano, *Defining the Atlantic Community*.

19. The original quote, "benevolent supremacy," comes from Charles Hilliard's *The Cross, the Sword, and the Dollar* (New York: North River Press, 1951). This concept was explored and further developed by McAlister, *Epic Encounters*; see esp. 43–83. The quote comes from Melani McAlister, "'Benevolent Supremacy': Biblical Epic Films, Suez, and the Cultural Politics of U.S. Power" (working paper, Yale University, September 2002), 204–5, http://opus.macmillan.yale.edu/workpaper/pdfs/MESV5-6.pdf.

20. Luce, "The American Century." Also see Lundestad, "'Empire by Invitation' in the American Century"; White, "The 'American Century' in World History"; and Herzstein, *Henry R. Luce, Time, and the American Crusade in Asia*.

21. Kozol, *Life's America*, 6.

22. Advertisement for Hudson Sprayers in *Soap and Sanitary Chemicals*, June 1944.

23. Although he argues that the Japanese interpreted and constructed racial identities that reversed and confounded Western conventions, asserting that Japan belonged at the top of the racial hierarchy, Gerald Horne recognizes the centrality of race in the Pacific war. See Horne, *Race War*. For an analysis of U.S. racial constructions, see Dower, *War without Mercy*. For an analysis of race and U.S. foreign policy prior to the Second World War, see Hunt, *Ideology and U.S. Foreign Policy*.

24. "Japan," *Life*, September 18, 1944, 60.

25. "Total Insect War Urged," *Science News Letter*, January 6, 1945, 5; Russell, *War and Nature*, esp. "Total War," 95–118, and "Annihilation," 119–44.

26. Moeller, *Shooting War*, 232.

27. Advertisement for Associated Chemicals, *Soap and Sanitary Chemicals* 21 (December 1945).

28. Fred L. Soper, "International Health Work in the Americas," speech before the National Health Assembly, May 3, 1948, Fred L. Soper Papers, http://profiles.nlm.nih.gov/VV/B/B/D/F/_/vvbbdf.pdf.

29. Sallares, *Malaria and Rome*, 90–93; Becker et al., *Mosquitoes and Their Control*, 176–77.

30. Farley, *To Cast Out Disease*, 113–22. Also see Paul Putnam and Lewis Hackett, "An Appraisal of the Malaria Endemic in Protected and Comparison Areas of Sardinia in the Years 1925–34," *Journal of the National Malaria Society* 5 (1946): 12–37.

31. Regarding the construction of the eradication campaign, others have made this observation as well. See Brown, "Failure-as-Success," 125; and Marcus Hall, "Fighting Malaria in Sardinia: DDT, the Rockefeller Foundation, and Imperial Environmentalism" (paper presented at the 2004 Fellows Conference, Environmental Values Project, Carnegie Council on Ethics and International Affairs).

32. Stapleton, "Internationalism and Nationalism," 127–34; and Bruce-Chwatt and Zulueta, *The Rise and Fall of Malaria in Europe*, esp. chap. 9, "Malaria in Italy." Postwar efforts to eradicate or control disease-bearing mosquitoes and other insect pests with DDT were global. Spray programs were instituted in Venezuela, Bolivia, Peru, Colom-

bia, Brazil, Italy, Greece, Sri Lanka (Ceylon), India, Thailand, Pakistan, Taiwan, Iraq, Iran, and the Philippines. For a brief summary of these postwar projects, see Russell, *Man's Mastery of Malaria*, esp. 160–90.

33. See Brown, "Failure-as-Success."

34. Quoted in Marcus Hall, "Today Sardinia, Tomorrow the World: Killing Mosquitos," *Bardpolitik* 5 (Fall 2004), 24. Also see Hall, "Today Sardinia, Tomorrow the World: Malaria, the Rockefeller Foundation, and Mosquito Eradication," Rockefeller Foundation Research Report, http://www.rockarch.org/publications/resrep/hall.pdf. Quote also found in Farley, *To Cast Out Disease*, 147.

35. Logan, *Ente regionale per la lotta anti-anofelica in Sardegna*.

36. Hall, "Today Sardinia, Tomorrow the World," 26.

37. Quoted in Farley, *To Cast Out Disease*, 147.

38. "Victory in Sardinia," *New York Times*, July 16, 1950.

39. Special Events Recordings, "Report from the ECA," interview with Mr. James D. Zellerbach, recorded January 25, 1949, and interview with Ben Wilbur, recorded January 25, 1949, Public Works to Reparations, Office of the Director Subject Files (Central Files), 1948–57, Mission to Italy, Records of U.S. Foreign Assistance, 1948–61, Record Group 469, National Archives at College Park, Md. (hereafter NACP).

40. Minister to Leon M. Dayton, summer 1950, Sardinia to Taxation, Office of the Director Subject Files (Central Files), 1948–57, Mission to Italy, Records of U.S. Foreign Assistance, 1948–61, Record Group 469, NACP.

41. M. L. Dayton, chief, ECA Special Mission to Italy to Joseph M. McDaniel Jr., assistant to the president, Ford Foundation, May 31, 1951, Sardinia to Taxation, Office of the Director Subject Files (Central Files), 1948–57, Mission to Italy, Records of U.S. Foreign Assistance, 1948–61, Record Group 469, NACP; also see "Information Officer from ECA/Rome," ECA Telegram, March 21, 1950, Public Works to Reparations, Office of the Director Subject Files (Central Files), 1948–57, Mission to Italy, Records of U.S. Foreign Assistance, 1948–61, Record Group 469, NACP.

42. "An Evaluation of Three Years of International Cooperation through the Marshall Plan," speech by M. L. Dayton, chief, ERP Special Mission to Italy Before the Italian Chamber of Commerce for the Americas, Rome, March 31, 1951, Public Works to Reparations, Office of the Director Subject Files (Central Files), 1948–57, Mission to Italy, Records of U.S. Foreign Assistance, 1948–61, Record Group 469, NACP.

43. "Visto Da Sinistra" and "Visto Da Destra," *Candido: Settimanale del sabato* (Milan), October 29, 1950, n.p. I would like to thank Alberto and Carlotta Guareschi for providing this source. I would also like to thank Chiara De Santi for her translation help and for providing me with her collection of *Candido* issues. For other Italian press accounts of Dayton's speech, see "Italian Press Reaction to Mr. Dayton's Speech Before The American Chamber of Commerce for Italy," Genoa, Turin, and Milan, October 19–21, 1950, Public Works to Reparations, Office of the Director Subject Files (Central Files), 1948–57, Mission to Italy, Records of U.S. Foreign Assistance, 1948–61, Record Group 469, NACP.

44. Giovannino Guareschi, "Considerazioni amene," *Candido*, October 29, 1950.

45. Winthrop Sargeant, "Anti-Communist Funnyman," *Life*, November 10, 1952, 115, 116.

46. "L'Industria Italiana, I Parassiti e . . . il D. DayTon," *Candido*, November 12, 1950.

47. Lord John Boyd-Orr to John A. Logan, superintendent of ERLAAS, Cagliari, June 8, 1950, Sardinia to Taxation, Office of the Director Subject Files (Central Files), 1948–57, Mission to Italy, Records of U.S. Foreign Assistance, 1948–61, Record Group 469, NACP. It is important to recognize that Boyd-Orr used the term "eliminate" rather than "eradicate" to describe the successes of the Sardinian campaign. This usage choice reflects the fact that those looking at the project from the outside had a different, and perhaps more realistic, conception of success.

48. Jack Straus and Robert Weil Jr. to Lee Dayton, August 31, 1951, Public Relations, Office of the Director Subject Files, 1948–1957, Mission to Italy, Records of U.S. Foreign Assistance, Record Group 469, NACP. M. L. Dayton to Robert Weil Jr., October 1, 1951, Public Relations, Office of the Director Subject Files, 1948–1957, Mission to Italy, Records of U.S. Foreign Assistance, Record Group 469, NACP.

49. "Italy in New York," *New York Times*, September 15, 1951; also see "Italian Fair Here Opened by Mayor," *New York Times*, September 11, 1941, and "Abroad at Home," *Time*, September 17, 1951.

50. For some brief examples, see C. H. Hoffmann and J. P. Linduska, "Considerations of the Biological Effects of DDT," *Scientific Monthly* 69, no. 2 (August 1949), 104–14, which suggested that the "most critical need is for a better understanding of the hazards implied by possible cumulative action." Also see J. A. Jenemann, "DDT and the Public Interest," *Soap and Sanitary Chemicals* 22 (January 1946): 121–23.

51. Paul Putnam and Lewis Hackett, "An Appraisal of the Malaria Endemic in Protected and Comparison Areas of Sardinia in the Years 1925–34," *Journal of the National Malaria Society* 5 (1946): 12–37.

52. The text of Truman's speech was printed in the *New York Times*, June 29, 1950.

53. "Point Four to Urge Capitalist Farms," *New York Times*, January 8, 1951.

54. Paul E. Deutschman, "Success Story from Sardinia," *Harper's*, December 1951, 64–70.

55. Packard and Brown, "Malaria, *Miseria*, and Underpopulation in Sardinia."

56. "Adventure in Sardinia," directed by Peter Baylis, produced for ECA U.K. by Associated British Pathe, Ltd., London, NARA ref. copy: PPC 16mm, 1R; and "Isle of Hope," produced by the MSA Film Section, Paris, for MSA Washington. See "Strength for the Free World" (series) for more information. NARA ref. copy: VT.

57. Barsam, *Nonfiction Film*, 277–80; Fedunkiw, "Malaria Films"; and for a more critical discussion on the development and use of public health films, see Reagan, Tomes, and Treichler, *Medicine's Moving Pictures*, 1–16. On Marshall Plan films, see Heffelfinger, "Foreign Policy, Domestic Fiction."

58. From "Italy's Recovery and United States Aid," editorial, *Il Messaggero di Roma*, April 4, 1953, translated by U.S. Foreign Assistance Agency, Public Works to Repara-

tions, Office of the Director Subject Files (Central Files), 1948–57, Mission to Italy, Records of U.S. Foreign Assistance, 1948–61, Record Group 469, NACP.

59. "The Italian Economy as Americans See It," *Esteri*, March 15, 1953, translated by U.S. Foreign Assistance Agency, Public Works to Reparations, Office of the Director Subject Files (Central Files), 1948–57, Mission to Italy, Records of U.S. Foreign Assistance, 1948–61, Record Group 469, NACP.

60. "DDT in Sardinia: How American Leadership Freed the Island from the Grip of Malaria," *Fortune*, March 1953, 113–20.

61. Edward F. Knipling, "Present Status of Mosquito Resistance to Insecticides," *American Journal of Tropical Medicine and Hygiene* 1 (1952): 389–94.

Chapter 3

1. Telegram from the League of Agricultural Communities of Jalisco and the Federation of Small Property Holders to President Don Adolfo Ruiz Cortines, February 24, 1953, folder 26, box 4, series 3223, RG 1.1, RAC. The foundation was extremely pleased with the telegram, which, in essence, wholeheartedly embraced its agricultural program. One Rockefeller official simply wrote, "Splendid!" across the top of the telegram.

2. Moreno, *Yankee Don't Go Home!*, 11, 45–81. Also see Schuler, *Mexico Between Hitler and Roosevelt*, 173–97; Kleinman, *A World of Hope, a World of Fear*, 121–23; White and Maze, *Henry A. Wallace*, 149–53; and Culver and Hyde, *American Dreamer*, 246–51.

3. Francis Sill Wickware, "Mexico's President: Avila Camacho Modifies Social Revolution of Last Six Years," *Life*, December 2, 1940, 96–105.

4. Henry A. Wallace, "Wallace in Mexico," *Wallace's Farmer*, February 22, 1941, 11.

5. George Messersmith to Secretary of State Cordell Hull, September 23, 1942, quoted in Perkins, *Geopolitics and the Green Revolution*, 126.

6. Memo from Raymond Fosdick, February 3, 1941, folder 70, box 11, series 323, RG 1.1, RAC.

7. Perkins, *Geopolitics and the Green Revolution*, 112, 114. Also see Deborah Fitzgerald, "Exporting American Agriculture: The Rockefeller Foundation in Mexico, 1943–1953," in Cueto, *Missionaries of Science*; Joseph Cotter, "The Mexican Agricultural Project," in Cueto, *Missionaries of Science*; Cotter, *Troubled Harvest*; Jennings, *Foundations of International Agricultural Research*; Sanderson, *The Transformation of Mexican Agriculture*; and Hewitt de Alcantara, *Modernizing Mexican Agriculture*.

8. Avila Camacho, quoted in Joseph Cotter, "The Mexican Agricultural Project," in Cueto, *Missionaries of Science*.

9. See Camp, *Entrepreneurs and Politics in Twentieth-Century Mexico*, and Joseph, Rubenstein, and Zolov, *Fragments of a Golden Age*.

10. Similar arguments about science and state formation are made elsewhere. See McCook, *States of Nature*, 8. Also see Soluri, *Banana Cultures*, 3; Chris Sheppard, "Rockefeller Foundation and Agriculture in Peru," Rockefeller Foundation Research

Report, http://www.rockarch.org/publications/resrep/shepherd.pdf; and, for more on science and development, see McBride *Missions for Science*.

11. Wright, *The Death of Ramón González*; Wright, "Rethinking the Circle of Poison"; Murray, *Cultivating Crisis*.

12. Alumni Awards, University of Minnesota, College of Food, Agricultural and Natural Resources Sciences, "Borlaug Through the Years," http://www.cfans.umn.edu/ AlumniFriends/awards/borlaug/borlaugthroughyears/index.htm.

13. Stakman, Bradfield, and Mangelsdorf, *Campaigns Against Hunger*, 33. In 1967, Stakman, Bradfield, and Mangelsdorf coauthored this book on their experience with MAP. The authors also listed the "improvement of soil management and tillage practices," in addition to the "introduction or development of better breeds of domestic animals and poultry, as well as better feeding methods and disease control" (ibid., 32–33). For more on Paul Mangelsdorf, see Peterson and Bianchi, *Maize Genetics and Breeding in the 20th Century*, 149–58.

14. See Birn, *Marriage of Convenience*.

15. Wilbur Downs to Hugh Smith, June 28, 1948, folder 151, box 18, series 1.1, RG 1.1, RAC.

16. Cueto, *Cold War, Deadly Fevers*, 128.

17. Stakman, "Oral History," 932.

18. According to the Rockefeller Foundation's annual report from 1944, only "$147,800 was appropriated for general expenses in 1944 and 1945, and $45,000 for the construction and equipment of an experimental laboratory." See *Rockefeller Foundation Annual Report*, 1944.

19. "Various Programs of World Planned by the Office of Special Studies for 1949," folder 59, box, 10, series 323, RG 1.2, RAC.

20. Deborah Fitzgerald, "Exporting American Agriculture," in Cueto, *Missionaries of Science*, 72–96.

21. Stakman, "Oral History," 1001, 1088. In her study of the Rockefeller Foundation's hookworm campaign in Mexico, Anne-Emmanuelle Birn suggested that the project established an unequal partnership between the foundation and the Mexican government, emphasizing that the foundation "insisted that Mexico request assistance," which, she argued, "immediately turned the Mexican government into a subservient player." Additionally, "the 'invitation' by the Mexicans," Birn continued, "obliged the acceptance of a lengthy list of Rockefeller requirements," including virtual independence in carrying out the program. Birn, "Public Health or Public Menace?"

22. Mexico Agricultural Program—Annual Report, 1943, folder 21, box 2, series 1.1, RG 6.13, RAC.

23. See Hurt, *American Agriculture*, 300.

24. Worster, *Dust Bowl*.

25. Interview with Norman Borlaug at http://www.livinghistoryfarm.org/podcasts/ BorlaugMexico.m4v.

26. Harrar to F. B. Hanson, January 5, 1944, folder 7, box 1, series 323, RG 1.1, RAC.

27. See Worster, "Transformations of the Earth," and Worster, *The Wealth of Nature*, 45–63.

28. Tentative Program of the Oficina de Estudios Especiales, S.A.G., for the Year 1950, folder 504, box 44, series 1.1, RG 6.13, RAC.

29. Report of the Oficina de Estudios Especiales, S.A.G., February 1, 1943–June 1, 1945, inclusive, folder 522, box 46, series 1.1, RG 6.13, RAC.

30. Informe de Los Progresos de la Oficinia de Estudios Especiales, July 1944 to December 1945, folder 523, box 46, series 1.1, RG 6.13, RAC.

31. Report of the Oficina de Estudios Especiales, June 1, 1945–June 1, 1946, inclusive, folder 523, box 46, series 1.1, RG 6.13, RAC.

32. Report of the Oficina de Estudios Especiales, June 1, 1945–June 1, 1946, inclusive, folder 523, box 46, series 1.1, RG 6.13, RAC.

33. "Various Programs of Work Planned by the Office of Special Studies for 1949," folder 59, box 10, series 323, RG 1.1, RAC.

34. See McWilliams, *American Pests*, esp. 144–93.

35. Simonian, *Defending the Land of the Jaguar*, 118–22.

36. Vogt, *El Hombre y la Tierra*, 55–65. See Perkins, *Geopolitics and the Green Revolution*, 136–37.

37. Vogt, *The Road to Survival*, 78, 191.

38. See Perkins, *Geopolitics and the Green Revolution*, 136–37, and Cullather, *The Hungry World*, 66.

39. Osborne, *Our Plundered Planet*, 9, 14. Also see Gottlieb, *Forcing the Spring*, 71–72.

40. "The Americas and the War," *Bulletin of the Pan American Union* 77, no. 9 (1943): 511.

41. Downs to Smith, March 27, 1947, folder 150, box 18, series 323, RG 1.1, RAC.

42. Downs to Smith, August 24, 1946, folder 2223, box 329, series 200, RG 2,RAC; Downs to Gahan, July 10, 1947, folder 50, box 18, series 323I, RG 1.1, RAC, Downs to Smith, March 19, 1947, folder 150, box 18, series 323, RG 1.1, RAC.

43. Downs to Smith, November 13, 1948, folder 151, box 18, series 323I, RG 1.1, RAC.

44. Downs to Smith, April 9, 1948, folder 151, box 18, series 323I, RG 1.1, RAC.

45. This was not the first time the United States intervened in Mexican foreign policy to ward off a "national threat." See Spenser, *The Impossible Triangle*.

46. Downs to Smith, September 11, 1948, meeting of the Malaria Council; present were Benitez Armas, Galo Soberón, Bustos, Vargas, Diaz Coller, Ing Ortega, Ing Morales and Downs. September 10, folder 151, box 18, series 323I, RG 1.1, RAC.

47. Alvin J. Cox, "Resume of DDT for Agriculture," *Agricultural Chemicals* (January 1947): 22–25; Alvin J. Cox, "A Resume of DDT, Insecticides for Agriculture," *Agricultural Chemicals* (February 1947): 29–35, 62–63; Milton Charles Winternitz, "DDT," in Weaver, *The Scientists Speak*, 88–92. From 1932 to 1955, Weaver was the director of the Rockefeller Foundation's Natural Science Division, which directed MAP.

48. Review of the Work of the Oficina de Estudios Especiales, S.A.G., for the year 1946–1947, folder 21, box 2, series 1.1, RG 6.13, RAC.

49. "Agricultural Suggestions," folder 13, box 2, series 3232, RG 1.1, RAC.

50. J. G. Harrar, "Fertilizer, Pesticide Use in Mexico," *Agricultural Chemicals* (February 1955): 28, 139.

51. See Annual Report of the Oficina de Estudios Especiales, S.A.G., September 1, 1952–August 31, 1953, folder 69, box 11, series 323, RG 1.1, RAC; Various Programs of Work Planned by the Office of Special Studies for 1949, folder 59, box 10, series 323, RG 1.2, RAC.

52. Fletcher, *The Pest War*, 45–63.

53. Frank B. Hanson to E. C. Stakman, n.d., folder 9, box 2, series 323, RG 1.1, RAC.

54. Annual Report of the Oficina de Estudios Especiales, S.A.G., September 1, 1952–August 31, 1953, folder 69, box 11, series 323, RG 1.1 RAC.

55. A report of the research conducted in entomology by the Oficina de Estudios Especiales, S.A.G., during the year 1951–1952, p. 2, folder 21, box 2, series 1.1, RG 6.13, RAC.

56. A report of the research conducted in entomology by the Oficina de Estudios Especiales, S.A.G., during the year 1953–1954, p. 3, folder 482, box 43, series 1.1, RG 6.13, RAC.

57. Mangelsdorf, report on a trip to Mexico for the Rockefeller Foundation, February/March 1946, folder 61, box 10, series 323, RG 1.2, RAC.

58. See Wright, "Rethinking the Circle of Poison," 47–49, and Bull, *A Growing Problem*; Weir and Shapiro, *The Circle of Poison*; and Murray, *Cultivating Crisis*.

59. Stakman, "Oral History," 1179.

60. Alvin J. Cox, "A Resume of DDT, Insecticides for Agriculture," *Agricultural Chemicals* (February 1947): 22–25.

61. *Rockefeller Foundation Annual Report*, 1962.

62. "Sección agrícola a cargo de los técnicos de Cotaxtla," *El Dictamen* (Vera Cruz), September 26, 1957.

63. OEE Annual Report, 1961–1962, folder 21, box 2, series 1.1, RG 6.13, RAC.

64. Douglas Barnes and William R. Young, "Insecticidas: Uselos Correctamente y Obtendra Buenos Resultados," Direccion General de Agricultura and the Oficina de Estudios Especiales, July 1965, folder 95, box 16, series 323, RG 1.2, RAC.

65. Deborah Fitzgerald, "Exporting American Agriculture," in Cueto, *Missionaries of Science*, 85. Since these bulletins were produced in Spanish, many smaller indigenous groups did not "benefit" from the foundation's expertise.

66. "En los Terrenos de la Escuela Nacional de Agricultura se Efectuaron Pruebas del Buevo Fumigante 'D-D,'" *Excelsior*, October 26, 1946, folder 13, box 2, series 323, RG 1.1, RAC.

67. J. George Harrar diary, May 31, 1951, RAC.

68. J. George Harrar diary, August 12, 1953, RAC.

69. Brandt, *Growth Company*, 458. DuPont eventually moved to Mexico in 1962. Dow Chemical was one of the earliest U.S. chemical companies to arrive, opening up offices in Mexico in 1954.

70. Garcia-Johnson, *Exporting Environmentalism*, 55 and 111.

71. Report of the Combined Mexican Working Party, The Major Long-term Trends in the Mexican Economy, with Particular Reference to Mexico's Capacity to Absorb Additional Foreign Investments, September 1952, Section Z-2, 32–36, folder 26, box 4, series 323, RG 1.1, RAC.

72. Cotter, *Troubled Harvest*, 233.

73. Hewitt de Alcantara, *Modernizing Mexican Agriculture*, 83–85.

74. William Cobb, unpublished manuscript, "The Historical Backgrounds of the Mexican Agricultural Program," March 1956, folder 62, box 10, series 323, RG 1.2, RAC.

75. Report of the Conference of Food and Agricultural Organization, Fourth Session, Washington, D.C., November 15–19, 1948.

76. Minutes of the Board of Consultants for Agriculture of the Rockefeller Foundation, October 29 and 30, 1953, folder 279, box 25, series 1.1, RG 6.13, RAC. Also see Stakman, "International Problems in Plant Disease Control," *Proceedings of the American Philosophical Society* 91, no. 1 (February 25, 1947): 95–111.

77. Advisory Committee for Agricultural Activities, The World Food Problem, Agriculture, and the Rockefeller Foundation, June 21, 1951, folder 23, box 3, series 915, RG 3, RAC. Also see Perkins, *Geopolitics and the Green Revolution*, 138.

78. Perkins, *Geopolitics and the Green Revolution*, 119. Perkins employs the concept of population–national security theory, suggesting that the foundation's work in population studies and agricultural development provided the foundation with a knowledge base that could influence the direction of foreign aid.

79. "Mexico Reports Second Largest Wheat Crop in 1955," *Foreign Crops and Markets*, June 6, 1955.

80. Aubrey Adams, "Modernizing Agricultural Landscapes: The Rockefeller Foundation in Mexico, 1943–1961" (paper presented at the annual conference of the American Society for Environmental History, Portland, Ore., March 10–14, 2010).

81. Norman E. Borlaug, "The Impact of Agricultural Research on Mexican Wheat Production," p. 11, folder 580, box 50, series 1.1, RG 6.13, RAC.

82. J. M. Ripley to William Cobb, March 15, 1955, folder 1, box 1, series 323, RG 1.2, RAC.

83. J. M. Ripley to William Cobb, January 14, 1957, folder 5, box 1, series 323, RG 1.2, RAC.

84. Harry Truman, speech delivered at the opening session of the Centennial Celebrations of the AAAS, September 13, 1948. Published in *Science* 108 (September 24, 1948): 313–14.

85. See, for example, Berman, *The Ideology of Philanthropy*.

86. Notes from George Harrar's diary, July 5, 1951, folder 45, box 18, series 1.1., RG 12.2. Also see letter from Stanley Andrews, administrator, Technical Cooperation Administrations, Executive Office of the President, Office of the Director for Mutual Security, to Dean Rusk, President of Rockefeller Foundation, June 10, 1953, folder 157, box 25, series 200, RG 2, RAC. Throughout this period, the State Department inquired if the Rockefeller Foundation would be willing to participate in President Truman's

Point Four Program of Technical Assistance to Developing Nations. While the foundation rejected the invitation, it is clear that MAP influenced the technical assistance program. See Rivas, *Missionary Capitalist*, and Perkins, *Geopolitics and the Green Revolution*.

87. J. George Harrar diary, March 19, 1953, folder 157, box 25, series 200, RG 2, RAC.

88. "Point Four to Urge Capitalist Farms," *New York Times*, January 8, 1951.

89. Simonian, *Defending the Land of the Jaguar*, 121. Also see Poleman, *The Papaloapan Project*, 3–123.

90. Telegram to Don Adolfo Ruiz Cortines from the League of Agricultural Communities of Jalisco and the Federation of Small Property Holders, February 24, 1953. Translated by MAP officials. Folder 26, box 4, series 323, RG 1.1, RAC.

91. See Cotter, *Troubled Harvest*. Within the United States, one of the more outspoken critics of the Rockefeller Foundation's program was Carl Sauer. Sauer was critical of the premise that industrial agriculture could be imposed on nations without an upheaval of local cultural, social, and economic systems.

92. González, *Zapotec Science*, 145.

93. "Problems & Progress," *Time*, September 12, 1955.

94. Stakman, "Oral History," 1106.

Chapter 4

1. Mumford, *Technics and Civilization*, 217.

2. Lewis Mumford to Jane Jacobs, July 22, 1958, Special Collections, John J. Burns Library, Boston College.

3. Fred Soper, "Eradication: 'To Pluck Up by the Roots,'" keynote address given to the Wisconsin Anti-Tuberculosis Association, published in the association's newsletter, *The Crusader*, June 1962, p. 4, http://profiles.nlm.nih.gov/VV/V/B/D/M/_?vvbbdm.pdf.

4. G. Macdonald, "The Theory of the Eradication of Malaria," World Health Organization, Malaria Division, WHO/Mal/173, April 27, 1956. All World Health Organization documents cited here can be found at http://www.who.int.

5. Le Prince, *Mosquito Control in Panama*, 291.

6. "LePrince, Malaria Fighter," *Public Health Reports* 71, no. 8 (August 1956): 756–58. Also see Bruce-Chwatt, "Ronald Ross, William Gorgas, and Malaria Eradication."

7. Sutter, "Nature's Agents or Agents of Empire?"

8. Patterson, *The Mosquito Wars*, 36, 42.

9. Mitchell, *The Amazon Journal of Roger Casement*, 60–73, 256; Smith, *Explorers of the Amazon*, 285–324.

10. Hardenburg, *The Putumayo—The Devil's Paradise—Travels in the Peruvian Amazon Region*, 37.

11. Hardenburg, *Mosquito Eradication*, 15.

12. Ibid., 51.

13. Ibid., 62.

14. Ibid., v.

15. Ibid., 99, 172, 158, and 160.

16. Malcolm Gladwell, "The Mosquito Killer," *New Yorker*, July 2, 2001.

17. Milton M. Levenson, "World Disease Control Imperiled by DDT Shortage and High Price," *New York Times*, February 18, 1951.

18. Jackson, "Cognition and Global Malaria Eradication Programme," 193–216.

19. Paul F. Russell, "International Preventative Medicine," *Scientific Monthly* 71, no. 6 (December 1950): 393–400.

20. Paul F. Russell was one of the leading Rockefeller Foundation malariologists who advocated for a much broader strategy, using DDT only as one of many tools in the fight against malaria. Downs to Smith, January 21, 1949, folder 152, box 18, series 323I, RG 1.1, RAC.

21. Russell, *Man's Mastery of Malaria*, 160.

22. Staples, *The Birth of Development*, 161.

23. Extract from a document on malaria eradication submitted by the Director-General to the Eighth World Health Assembly, A8/P&B/10, May 3, 1955.

24. Staples, *The Birth of Development*, 166, 170. The U.S. appropriated $27.2 million, $26.2 million, and $32.0 million during 1958, 1959, and 1960, respectively, to support the worldwide malaria-eradication program. See Eastern Mediterranean Region World Health Organization Report on the Second Regional Conference on Malaria Eradication, Addis Ababa, November 16–21, 1959, WHO/Mal/265, May 20, 1959, 24.

25. Members of the Rockefeller Foundation public health campaign in Mexico faced difficult hurdles in administering indoor spraying projects. Locals often refused to allow spray teams into their communities or homes because they did not see any merit in the public health campaign. This occasionally led to violent confrontations. See Cuerto, *Cold War, Deadly Fevers*, 128, and Payne to Smith, October 22, 1945, folder 149, box 18, series 313I, RG 1.1, RAC.

26. See Marcus Hall, "Fighting Malaria in Sardinia: DDT, the Rockefeller Foundation, and Imperial Environmentalism" (paper presented at the 2004 Fellows Conference, Environmental Values Project, Carnegie Council on Ethics and International Affairs).

27. Expert Committee on Malaria, report on Dr. Pampana's Mission to Greece and Italy, WHO.IC/Mal./8, August 21, 1947, 8.

28. Report of the Fourth Session, Kampala, Uganda, December 11–16, 1950, WHO/Mal/70, December 27, 1950, 12.

29. See Robert L. Metcalf, "Physiological Basis for Insect Resistance to Insecticides," *Physiological Reviews* 35 (1955): 197–223.

30. Dr. Alvaro Lozano Morales, director, National Institute Navalmoral de la Mata, and Dr. Juan Gil Collado, entomologist, member, WHO Expert Advisory Panel on Insecticides, "H-24, A New Process for the Reactivation of Insecticides Preliminary Trials. Action on Resistant Domestic Flies," WHO/Mal/124, March 11, 1955.

31. Expert Committee on Malaria, Seventh Report, Lisbon, September 15–23, 1958, WHO/Mal/210, October 24, 1958.

32. Fifth Summary of Cases of Insecticide Resistance and Records of Susceptibility Tests on Anopheline Mosquitoes, WHO/Mal/242, September 15, 1959.

33. R. W. Babione, PAHO/WHO, Washington, D.C., "Special Technical Problems in Malaria Eradication in Latin America," WHO/Mal/430, February 6, 1964, 1.

34. Mallet, "The Evolution of Insecticide Resistance." Also see Brogdon and McAllister, "Insecticide Resistance and Vector Control"; Georghioui, "The Evolution of Resistance."

35. Hoy, "Myths, Models and Mitigation of Resistance to Pesticides."

36. Dr. J. R. Busvine, "Insecticide-Resistant Strains," WHO/Mal/142, October 21, 1955; P. M. Bernard, "Médecin-Colonel," Malaria Control in Tropical Africa, WHO/ Mal/131, September 2, 1955, 4.

37. "Mosquito Versus Science," *World Health: The World United Against Malaria* special issue, 1955, 15.

38. Robert L. Metcalf, "Physiological Basis for Insect Resistance to Insecticides," *Physiological Reviews* 35 (1955): 228.

39. Mark F. Boyd, "Planning For Malaria Control," WHO.IC/Mal/20, April 22, 1948.

40. Expert Committee on Malaria, Report of the Fourth Session, WHO/Mal/70, January 17, 1951.

41. A. Lebrun and M. A. Ruzette, "What Results Can We Expect From the Use of Aerial Methods in Malaria Control?" WHO/Mal/156, February 15, 1956.

42. Clay Lyle, "Achievements and Possibilities in Pest Eradication," *Journal of Economic Entomology* 40 (February 1947): 1–8.

43. Hal Higdon, "Well, If not DDT, then What?" *New York Times Magazine*, January 11, 1970, 26–27, 48–58.

44. Buhs, *The Fire Ant Wars*, 38. Also see Daniel, *Toxic Drift*, 48–70, 159–70.

45. "Pesticides and Wildlife," *Journal of Agricultural and Food Chemistry* 7, no. 4 (April 1959): 236.

46. Tschinkel, *The Fire Ants*, 72.

47. C. Payne to Clifford Hardin, February 26, 1971, box 110, file 1, W. R. Poage Papers, Baylor University Collection of Political Materials, Waco, Tex. (hereafter BCPM).

48. See correspondence between Knipling and John Perkins, author of *Geopolitics and the Green Revolution*, box 11, folder 549, Edward Fred Knipling Papers, Special Collections of the National Agricultural Library, Beltsville, Md.

49. Daniel, *Toxic Drift*, 5.

50. A. V. Hill, "The Ethical Dilemma of Science," *Nature*, September 6, 1952, 388–93.

51. E. J. Pampana, "Unexpected Cost Increases of Malaria Eradication Programmes," WHO/Mal/208, August 26, 1958, 3.

52. Staples, *The Birth of Development*, 169.

53. Paul Russell to Lowell T. Coggeshall, January 15, 1959, folder 463, box 48, series 100, RG 1.1, RAC.

54. Dr. L. J. Bruce-Chwatt, "Malaria Research and Eradication in the USSR," WHO/ Mal/222, April 8, 1959, 4.

55. Médecin-Colonel P. M. Bernard, Malaria Control in Tropical Africa, WHO/ Mal/131, September 2, 1955.

56. Extract from a document on malaria eradication submitted by the Director-General to the Eighth World Health Assembly, A8/P&B/10, May 3, 1955.

57. Eastern Mediterranean Region World Health Organization Report on the Second Regional Conference on Malaria Eradication, Addis Ababa, November 16–21, 1959, WHO/Mal/265, May 20, 1959, 9.

58. Progress of the National Malaria Eradication Programme of India, WHO/ Mal/300, July 15, 1961.

59. Perkovich, India's Nuclear Bomb, 15.

60. C. W. Kearns, "The Mode of Action of Insecticides," Annual Review of Entomology 1 (1956): 123–48; E. P. W. Winteringham and S. E. Lewis, "On the Mode of Action of Insecticides," Annual Review of Entomology 4 (1959): 303–18.

61. Davies et al., "DDT, Pyrethrins, Pyrethroids and Insect Sodium Channels."

62. "Spray DDT in Polio Area," New York Times, August 14, 1945; "DDT Sprayed Over Rockford, Ill., in Test of Power to Halt Polio," New York Times, August 20, 1945; "Spraying District with DDT Thought Useless as Polio Gains," Washington Post, August 21, 1945.

63. See Evert W. Johnson, "Role of Aircraft in Forest-Pest Control," Scientific Monthly 79, no. 6 (December 1954): 379–91; and David C. Powell, "Douglas-Fir Tussock Moth in the Blue Mountains," http://www.fs.fed.us/r6/uma/publications/history/tussock%20moth%20story.pdf.

64. R. F. Addison and T. G. Smith, "Organochlorine Residue Levels in Arctic Ringed Seals: Variation with Age and Sex," Oikos 25, no. 3 (1974): 335–37; G. W. Bowes and C. J. Jonkel, "Presence and Distribution of Polychlorinated Biphenyls (PCB) in Arctic and Subarctic Marine Food Chains," Journal of the Fisheries Research Board of Canada 32, no. 11 (December 1976): 2111–12.

65. As Knipling demonstrated, not all eradication programs ended in failure. Smallpox eradication was another major health success story. See Manela, "A Pox on Your Narrative."

Chapter 5

1. United Nations Development Decade: A Programme for International Economic Co-operation 1710 (XVI), resolutions adopted on the reports of the second committee, December 19, 1961.

2. Latham, Modernization as Ideology, esp. 69–150.

3. Merrill, Bread and the Ballot, 170. Also see Rotter, Comrades at Odds.

4. For a rich and insightful history on the social, political, and environmental changes brought about by postwar development projects, see Shiva, The Violence of Green Revolution. Also see Brar, Green Revolution; Perkins, Geopolitics and the Green Revolution; Cullather, "Miracles of Modernization"; and Cullather, The Hungry World.

5. McWilliams, American Pests, 194.

6. Although the modern environmental movement has a longer history, I take the term "Green Revolution" from Kirkpatrick Sale's book, *The Green Revolution*, which erroneously places the beginning of American environmentalism at the publication of *Silent Spring*. Nevertheless, Sale offers an insightful critique of the movement.

7. Memo from Marie Rodell to editor, the *New Yorker* Records, c. 1924–1984, IV, Editorial Business 4.4, General Files, 1927–89, box 965, folder 4, New York Public Library. According to the *New Yorker*, seventy-one papers published news stories on Carson's book, sixty-one papers carried relevant editorials, thirty-one columnists mentioned the book, and thirty-three papers carried letters to the editor on the topic.

8. Carson, *Silent Spring*, 15, 5.

9. Ibid., 276.

10. Ibid., 8 (emphasis added).

11. Ibid.

12. Nash, *Inescapable Ecologies*, 157.

13. Carson, *Silent Spring*, 12.

14. "CBS Reports: The Silent Spring of Rachel Carson" as broadcast over the CBS Television Network, Wednesday April 3, 1963, 7:30–8:30 DST; Murphy, *What a Book Can Do*, 126. Murphy also indicates that, prior to the airing of the program, CBS received an inordinate amount of mail asking for "a fair and impartial discussion" of the pesticide issue. CBS producers suspected the letters were part of a chemical industry letter-writing campaign. See Murphy, *What a Book Can Do*, 115–16.

15. See Dunlap, *DDT*.

16. The focal point of the questioning concerning the adequate regulation of pesticides centered on the Federal Insecticide, Fungicide, and Rodenticide Act (FIFRA), passed in 1947, which legislated registration and labeling requirements. Also discussed was the Miller Pesticide Amendment (1954), which established tolerance levels on raw agricultural commodities.

17. Interagency Coordination in Environmental Hazards (Pesticides), Senate Committee on Government Operations: Coordination of Activities Relating to the Use of Pesticides, CIS-NO 88 S1595-3-A, 220.

18. Jennings, *Foundations of International Agricultural Research: Science and Politics in Mexican Agriculture*, 141–46.

19. Don Paarlberg, "Norman Borlaug—Hunger Fighter," Foreign Economic Development Service, U.S. Department of Agriculture, cooperating with U.S. Agency for International Development, December 1970.

20. Bickel, *Facing Starvation*, 260.

21. Bill Ganzel, "Norman Borlaug and Henry Beachell," http://www.livinghistory farm.org/farminginthe50s/crops_15.html; Perkins, *Geopolitics and the Green Revolution*, 236–37.

22. Cullather, "Miracles of Modernization."

23. Pearce, *Rostow, Kennedy, and the Rhetoric of Foreign Aid*, 118. Also see Gilman, *Mandarins of the Future*, 1–23.

24. On the consequences of high modernism, see Scott, *Seeing Like a State*.

25. Carson, *Silent Spring*, 297.

26. "Accentuating the Positive," *Chemical Week*, September 22, 1962, 107.

27. McGrath, *Scientists, Business, and the State, 1890–1960*, and Wang, *American Science in an Age of Anxiety*.

28. Still committed to military science, Bush nevertheless began to express reservations about how dominant the military-industrial complex had become. See Zachary, *Endless Frontier*, esp. 379–404.

29. See Russell, *War and Nature*, 184–203.

30. For an analysis of the gendered scientific debate that enveloped *Silent Spring*, see Hazlett, "'Woman vs. Man vs. Bugs.'"

31. "CBS Reports: "The Verdict on the Silent Spring of Rachel Carson" as broadcast over the CBS Television Network, Wednesday May 15, 1963, 7:30–8:30 DST. The report quotes *Time*'s characterization.

32. T. Swarbrick, "Carson Versus Chemicals," *The Grower*, February 9, 1963, 249, quoted in Groshong, "Noisy Response to *Silent Spring*," 38.

33. Edwin Diamond, "The Myth of the 'Pesticide Menace,'" *Saturday Evening Post*, September 28, 1962, 16.

34. Jon Lee, "Silent Spring is Now Noisy Summer," *New York Times*, July 22, 1962, 1.

35. Fredrick J. Stare, "Some Comments on *Silent Spring*," *Nutrition Reviews* 21, no. 1 (January 1963): 1–4; L. J. Teply, letter to the editor, *Chemical Week*, July 28, 1962, 6.

36. Lear, *Rachel Carson*, 206.

37. Jonathan Norton Leonard, "—And His Wonders in the Deep," *New York Times*, July 1, 1951.

38. Lear, *Rachel Carson*, 275.

39. Ibid., 276.

40. Hynes, *The Recurring Silent Spring*, 166.

41. "Pesticide Sales Pick Up," *Chemical and Engineering News*, July 2, 1962, 22.

42. Paul L. Kronick, letter to the editor, *Chemical and Engineering News*, July 16, 1962.

43. May, *Homeward Bound*, 27; Egan, *Barry Commoner and the Science of Survival*, 84.

44. Worster, *Nature's Economy*, 37.

45. See Kinkela, "The Ecological Landscapes of Jane Jacobs and Rachel Carson."

46. Thomas H. Jukes, "A Town in Harmony," *Chemical Week*, August 18, 1962, 5.

47. Ibid.

48. "Letters," *Chemical Week*, September 15, 1962, 7.

49. "The Desolate Year," *Monsanto Magazine*, October 1962, 5 (emphasis in original).

50. Ibid.

51. Bruce Benedict, "Rachel Carson Bugs the Entomologists," *San Francisco Chronicle*, June 29, 1962.

52. For an insightful discussion on the use of "horde" within the pesticide–Cold War context, see Russell, *War and Nature*, 187–89.

53. "Wanted: For Murder and Theft," *Cyanamid Magazine*, Winter 1963, 28.

54. I would like to thank Bill Marotti for his screening of the original Japanese version of *Godzilla*, *Gojira* (1954), directed by Ishirô Honda.

55. *Silent Spring's* publisher, Houghton Mifflin, completely supported Rachel Carson.

56. Louis A. McLean, Velsicol Chemical Corporation, to Mr. William E. Spaulding, Houghton-Mifflin Company, August 2, 1962, NYPL, Manuscripts and Archives Division, the *New Yorker* Records, Editorial Business, 4.4, General Files, 1927–89, box 965, folder 4.

57. "Accentuating the Positive," *Chemical Week*, September 22, 1962, 107.

58. *Fact and Fancy* was an extremely important industry document. When the American Medical Association commented on the *Silent Spring* controversy, it suggested that interested parties who wanted the "facts" should refer to the industry pamphlet. In one of her few public responses to the controversy surrounding her book, and to the AMA's recommendation, Rachel Carson stated in a speech to the Women's National Press Club: "Such a liaison between science and industry is a growing phenomenon. . . . The AMA, through its newspaper, has just referred physicians to a pesticide trade association for information to help them answer patients' questions about the effects of pesticides on man. I am sure physicians have a need for information on this subject. But I would like to see them referred to authoritative scientific or medical literature — not a trade organization whose business it is to promote the sales of pesticides. We see scientific societies acknowledging as 'sustaining associates' a dozen or more giants of a related industry. When the scientific organization speaks, whose voice do we hear — that of science? Or of the sustaining industry?" (Paul Brooks, *The House of Life: Rachel Carson at Work* [Boston: Houghton Mifflin, 1972], 303–4).

59. "Accentuating the Positive," *Chemical Week*, September 22, 1962, 107.

60. Ibid.

61. Murphy, *What a Book Can Do*, 117.

62. "Food: Will We Always Have Enough?" *Cyanamid Magazine*, Fall 1963, 3–7.

63. Robertson, "The Population Bomb," and Connelly, *Fatal Misconception*.

64. "Food: Will We Always Have Enough?" *Cyanamid Magazine*, Fall 1963, 3–7 (emphasis in original). "Nine bean rows in a bee-loud glade" was a reference to William Butler Yeats's poem "The Lake Isle of Innisfree": "I will arise and go now, and go to Innisfree, / And a small cabin build there, of clay and wattles made; / Nine bean rows will I have there, a hive for the honeybee, / And live alone in the bee-loud glade."

65. White House Central Files, Subject File, The White House Message on Foreign Aid, SP 2–3 1964, Foreign Aid, White House Central Files, Lyndon Baines Johnson Presidential Library, Austin, Tex. (hereafter WHCF LBJ Library).

66. White House Central Files, Subject File, U.S.A.I.D. Administrative History, p. 181, box 1, WHCF LBJ Library. Also see Doel and Harper, "Prometheus Unleashed."

67. Memorandum from Orville L. Freeman to President Johnson, March 22, 1966, WHCF, CM, box 4, LBJ Library.

68. White House Central Files, Subject File, W. W. Rostow to Mr. Frank P. Davidson, president, Technical Projects, Ltd., September 8, 1966, WHCF, CM, box 4, LBJ Library.

69. White House Central Files, Subject File, Statement By the President Upon Signing H.R. 14929—Food for Freedom Bill, WHCF, Legislation, box 167, LBJ Library.

70. White House Central Files, Subject File, Memorandum for Doug Cater from Donald F. Hornig, special assistant to the president for science and technology, October 14, 1966, WHCF, CM, box 4, LBJ Library. For more on the intersection of technology, modernity, and U.S. foreign policy, see Cullather, "Miracles of Modernization."

71. Rostow, *Eisenhower, Kennedy, and Foreign Aid*, 177.

72. "Administrative History of AID," 205. History of Agency for International Development, 2 vols. LBJ Library.

73. Posgate, "Fertilizers for India's Green Revolution"; Barker, Herdt, and Rose, *The Rice Economy of Asia*, 75 and 251. Also see Clevo Wilson, "Environmental and Human Costs of Commercial Agricultural Production in South Asia."

74. Cooper and Packard, *International Development and the Social Sciences*, 19.

75. Shiva, *The Violence of the Green Revolution*.

76. Draft paper by Orville Freeman and Lester Brown, "New Hope for a Hungry World," WHCF, CM, box 4, LBJ Library.

77. Graham, *Since Silent Spring*, 176–77.

78. Whitten served in Congress from 1941 to 1995, making him, at the end of his career, the longest-serving congressman in the history of the United States. William Ruckelshaus, who in 1970 would become the first administrator of the Environmental Protection Agency, remembered his first encounter with the powerful congressman: "The first time I met him he gave me an autographed copy of his book which . . . was *very* antagonistic to any regulation of pesticides. He believed it was all a lot of nonsense." William D. Ruckelshaus, interview by Dr. Michael Gorn, January 1993, Oral History Interview, Environmental Protection Agency, EPA 202-k-92-003, http://www.epa.gov/history/publications/print/ruck.htm.

79. Whitten, *That We May Live*, 214.

80. McMillen, *Bugs or People?*, 29.

81. Ibid., 115.

82. Tom Edwards, "Life through a Spray Nozzle? What's Behind Concern over Bug Chemical? Anti-Pest Program Sprays Questioned," newspaper unknown (Illinois), July 5, 1962, *New Yorker* Collection, box 1292, folder 11, New York Public Library.

83. Ed Roth to President Johnson, Mr. and Mrs. J. Buchwald to President Johnson, and M. A. Cheney to President Johnson, "Public Outcry of the 'Hunger in America,'" CBS Report, May 21, 1968, WHCF, CM, box 4, LBJ Library.

84. For example, see Carson, *Silent Spring*, 46, 64, 118.

85. There is a great deal of literature on American nature writers and the concept of pristine wilderness. Perhaps the most influential is still Nash, *Wilderness and the American Mind*.

86. See Schneider, *The Age of Romanticism*, esp. chap. 6, "Romantic Legacies: Nationalism and Environmentalism." Also see Lytle, *The Gentle Subversive*.

Chapter 6

1. Bruce-Chwatt, "Malaria Eradication at the Crossroads"; Staples, *The Birth of Development*, 179.

2. Phillip Shabecoff, "A Decade of Disappointment for Asia," *New York Times*, January 19, 1970; "The Green Revolution Turns Brown," *Washington Post*, February 19, 1972; and Lester R. Brown, "The Somber World Food Outlook," *Wall Street Journal*, March 26, 1973.

3. The founding members of the organization were Dennis Puleston, Carol Yannacone, H. Lewis Bates Jr., Robert Burnap, Arthur Cooley, Robert E. Smolker, and Anthony S. Taormina. Much work has been done on the Environmental Defense Fund and its work to ban DDT domestically. For an excellent reading of the EDF and DDT, see Dunlap, *DDT*. Dunlap's study details the legal strategies against the domestic ban of DDT. While his study does a masterful job narrating the political and legal issues surrounding the DDT debate, international or transnational issues are not discussed. For a detailed summary of the Wisconsin hearing, see Henkin, Merta, and Staples, *The Environment, the Establishment, and the Law*; and Orie L. Loucks, "The Trial of DDT in Wisconsin," in Harte and Socolow, *Patient Earth*, 88–111. For a brief institutional history of the EDF in the 1970s and 1980s, see Taylor, *Ahead of the Curve*.

4. Francis X. Clines, "Conservationists Press Fight Against Pesticides," *New York Times*, November 26, 1967.

5. Hal Higdon, "Obituary for DDT (in Michigan)," *New York Times Magazine*, July 6, 1969.

6. Loren H. Osman, "DDT 'Trial' Opens Monday in Midwest," *Washington Post*, December 1, 1968.

7. Scheffer, *The Shaping of Environmentalism in America*, 136.

8. Amyas Ames, "A Spokesman for the Interests of Man," (manuscript) Office Files, Lee Rogers, box 11, number 413, Environmental Defense Fund Archives, Manuscript Collection, 232, SUNY Stony Brook, Stony Brook, N.Y. (hereafter EDF Archives).

9. Frank Graham Jr., "Taking Polluters to Court," *New Republic*, January 13, 1968.

10. African Americans' use of the court system during the postwar period led to landmark civil rights cases, most notably *Brown v. Board of Education*.

11. "EDF: New Threat to Pesticides," *Farm Chemicals*, January 1968, 34.

12. Gordon L. Berg, "A Crack in Our Scientific Armor," *Farm Chemicals*, February 1968, 142.

13. Ibid., 36.

14. "Wisconsin is Site of Next Round in Fight to Ban DDT," *Washington Post*, April 26, 1969.

15. Berg, "A Crack in Our Scientific Armor."

16. Dunlap, *DDT*, 150, 157–58.

17. Loren H. Osman, "DDT 'Trial' Opens Monday in Midwest," *Washington Post*, December 1, 1968.

18. Richard D. James, "Periled Pesticide: Wisconsin Hearing on Bid to Ban DDT Could Affect Future of All Such Products," *Wall Street Journal*, March 4, 1969.

19. Charles F. Wurster, "DDT Goes to Trial in Madison," *Bioscience* 19, no. 9 (September 1969), 809–13.

20. Remarks by Senator Gaylord Nelson before the Department of Natural Resources: Petition of Citizens Natural Resources Association Inc. and Wisconsin Division, Izaak Walton, League of America Inc., for a declaratory ruling on use of dichloro-diphenyl-trichloro-ethane, commonly known as DDT, in the state of Wisconsin, Monday, December 2, 1968, Madison, Wisconsin, 24, EDF Archives (hereafter Wisconsin DDT hearing).

21. Wisconsin DDT hearing. Also see Robert W. Risebrough, P. Rieche, D. B. Peakall, S. G. Herman, and M. N. Kirven, "Polychlorinated Biphenyls in the Global Ecosystem," *Nature* 220 (December 14, 1968): 1098–1102. Also see Robert W. Risebrough et al., "DDT Residues in Pacific Sea Birds: A Persistent Insecticide in Marine Food Chains," *Nature* 216 (November 11, 1967): 589–91.

22. Wisconsin DDT hearing, 956.

23. Ibid., 961.

24. Ibid., 962.

25. Higdon, "Obituary for DDT"; "American Cyanamid Co. Cuts Price of Cythion, A Malathion Insecticide," *Wall Street Journal*, November 8, 1968.

26. The terms "cancellation" and "suspension" were defined in the federal laws regulating pesticides. Both terms acknowledge that a review process is under way. Suspension requires all uses of the pesticide in question to cease during the review process. Cancellation means the pesticide in question can still be used during the review process. Manufacturers may appeal suspension or cancellation notices, which could lead to a public hearing, if one is deemed necessary. For more, see *Audubon* 75, no. 6 (November 1973): 121–24.

27. "Studies Show DDT To Be Carcinogenic: Petition to FDA Requests Zero Tolerance in Foods," October 1, 1969, EDF Archives.

28. Dunlap, *DDT*, 206.

29. Bosso, *Pesticides and Politics*, 138, 126–27.

30. Ibid., 140.

31. Secretary's Commission on Pesticides and Their Relationship to Environmental Health, Recommendation and Summaries, Department of Health, Education, and Welfare, November 1969, 21.

32. Dunlap, *DDT*, 203.

33. Marti Mueller, "HEW Examines Cancer Institute Report," *Science* 164 (June 27, 1969): 1503.

34. "Wide Curb on DDT Scheduled by U.S. in Next 2 Years," *New York Times*, November 13, 1969.

35. Richard D. Lyons, "Hickels Extends Pesticide Curb," *New York Times*, June 18, 1970.

36. Robert Rodale, "The Wisconsin DDT Story in Picture and Words," *Organic Gardening and Farming*, July 1969.

37. Peter Hough, "Return to Sender: Regulating the International Trade in Pesticides" (paper presented at the ECPR workshop, The Politics of Food, Copenhagen, April 14–19, 2000).

38. *Environmental Defense Fund Annual Report*, 1971, EDF Archives.

39. McGeorge Bundy to Congressman William Poage, April 12, 1971, box 110, file 4, Poage Papers, BCPM.

40. Sponsors of Science, DDT fundraising letter, box 111, file 3, Poage Papers, BCPM. All correspondence and fundraising letters were signed by Betty Chapman, from Madison, Wisconsin, who was most likely married to the former professor of entomology from the University of Wisconsin, Madison, Keith Chapman. Both Chapmans have passed away. The Distinguished Graduate Fellowship in Entomology has been created in their name.

41. Betty Chapman to Dr. F. W. Whittemore, Food and Agriculture Organization, May 27, 1971, box 110, file 5, Poage Papers, BCPM.

42. Betty Chapman to Representative Harley Staggers, October 12, 1971, box 111, file 3, Poage Papers, BCPM.

43. Norman E. Borlaug to Miss Betty Chapman, October 4, 1971, box 111, file 3, Poage Papers, BCPM.

44. Widely published and at the forefront of the DDT controversy, Wurster formed much of the pesticide policy of the EDF. His articles appeared in a wide range of publications, including *Science, Field and Stream, Smithsonian, Innovation*, and the *New York Times*. For example, see George M. Woodwell, Charles F. Wurster, and Peter A. Isaacson, "DDT Residues in an East Coast Estuary: A Case of Biological Concentration of a Persistent Insecticide," *Science* 156, no. 3776 (May 12, 1967): 821–24.

45. Henkin, Merta, and Staples, *The Environment, the Establishment, and the Law*, 27–43. Chapter 3 pertains to Wurster's direct examination.

46. "DDT: Universal Contamination of the Earth?" *Roche Medical Image and Commentary*, March 1970, 14–18.

47. Charles F. Wurster, "Its Persistency Threatens Disaster to Beast and Man," *Smithsonian*, October 1970, 49.

48. Charles F. Wurster, "DDT Makes Matters Worse," *New York Times*, August 12, 1971.

49. The text of this testimony has caused much controversy. Many social critics who condemn environmentalists for their position on DDT have used this speech to suggest that the EDF, and environmentalists en masse, were grossly indifferent to the plight of people overseas. See, for example, Whelan, *Toxic Terror*, and Ray, *Environmental Overkill*. After a search of the hearing records, efforts to find the original speech have been unsuccessful. My efforts here are to provide a historic context for the DDT controversy, especially regarding the international issues that have remained political questions for so long. Also see Oreskes and Conway, *Merchants of Doubt*, 216–40.

50. Hearing, Committee on Agriculture, House of Representatives, February and March 1971, CIS NO 71-H161-3, 266–68.

51. Thomas Jukes to Robert Poage, June 30, 1972, box 112, file 1, Poage Papers, BCPM.

52. Historians have begun to examine the complicated history of race during the 1970s, especially as it pertains to ideas of whiteness. See Lassiter, *The Silent Majority*; Kruse, *White Flight*; and Thomas J. Sugrue and John D. Skrentny, "The White Ethic Strategy," in Schulman and Zelizer, *Rightward Bound*.

53. Hearing, Committee on Agriculture, House of Representatives, February and March 1971, CIS NO 71-H161-3, 266–68.

54. Thanks to the many archivists at the National Archives and Library of Congress who tried to locate these committee records.

55. See note 49 above. More recently, see Paul Driessen's praise for the film *Three Billion and Counting*, which again paints Wurster in a negative light. "Malaria: 3 Billion and Counting," *Family Security Matters*, September 13, 2010, http://www.familysecuritymatters.org/publications/id.7342/pub_detail.asp.

56. See Bruce-Chwatt, "Malaria Eradication at the Crossroads." Consolidation was the third of four phases of the eradication program. Coming after the attack phase, consolidation was designed to survey trouble spots and quickly extinguish any recurrence of the disease.

57. Fogleman, *Guide to the National Environmental Policy Act*, 3.

58. William D. Ruckelshaus, interviewed by Dr. Michael Gorn, January 1993, Oral History Interview, Environmental Protection Agency, EPA 202-K-92-003, http://www.epa.gov/history/publications/print/ruck.htm.

59. E. W. Kenworthy, "U.S. Court Orders New Unit to File Notice of DDT Ban," *New York Times*, January 8, 1971.

60. Roberta Hornig, "Ruckelshaus Curbs Shipments of DDT," *Evening Star*, January 15, 1971.

61. See Robert Gillette, "DDT: On Field and Courtroom a Persistent Pesticide Lives On," *Science* 174 (1971): 1108–10. Robert L. Ackerly and Charles A. O'Connor represented the petitioners—the formulators and manufacturers. Elliot C. Metcalfe and Raymond Fullerton represented the interests of the USDA. William A. Butler, the EDF's lead attorney, represented the interests of the EDF, the Sierra Club, the National Audubon Society, and the West Michigan Environmental Action Council. Blaine Fielding argued the case on behalf of the EPA.

62. Despite the dramatic change for assigning safety requirements, the FIFRA amendment of 1964 passed with "little controversy." See Bosso, *Pesticides and Politics*, 126.

63. Environmental Protection Agency, Consolidated DDT Hearings, 1–3, Cancellation Hearing files, EDF Archives (hereafter Consolidated DDT Hearings).

64. For an account of the Consolidated DDT Hearings that focuses on the overall strategies of the petitioners and interveners as well as Examiner Sweeney's handling of the case, see Dunlap, *DDT*, 211–30.

65. Agriculture Secretary Harding was one of the principal advocates for the voluntary withdrawal of DDT. Harding, however, made no connection between the environmental movement's campaign against DDT and its voluntary reduction. See Jane E. Brody, "Use of DDT at a 20-Year Low, Chiefly Due to Voluntary Action," *New York Times*, July 20, 1970.

66. Mary S. Wolff, Paolo G. Toniolo, et al., "Blood Levels of Organochlorine Residues and Risk of Breast Cancer," *Journal of the National Cancer Institute* 85 (1993): 648–52; Barbara A. Cohn, Mary S. Wolff, Piera M. Cirillo, and Robert I. Sholtz, "DDT and Breast Cancer in Young Women: New Data on the Significance of Age at Exposure," *Environmental Health Perspectives* 115, no. 10 (October 2007): 1406–14.

67. For a short overview of research on endocrine disruptors, see David Crews, Emily Willingham, and James K. Skipper, "Endocrine Disruptors: Present Issues, Future Directions," *Quarterly Review of Biology* 75, no. 3 (September 2000): 243–60. For a historical analysis, see Nancy Langston, "The Retreat From Precaution: Regulating Diethylstilbestrol (DES), Endocrine Disruptors, and Environmental Health," *Environmental History* 13, no. 1 (2008): 41–65.

68. Consolidated DDT Hearings, 15.

69. Ibid., 15–16.

70. Letter to Knipling, box 10, folder 449, Edward Fred Knipling Papers, Special Collections of the National Agricultural Library, Beltsville, Md.

71. Hal Higdon, "Well, If Not DDT, Then What?" *New York Times Magazine*, January 11, 1970, 26–27, 48–58.

72. Norman E. Borlaug, "Where Ecologists Go Wrong," *The Sun*, November 14, 1971, box 124, file 827, Hyde Murray Papers, BCPM.

73. David Kirkman, "Expert Assails Popular Notions about Ecology," October 11, 1971. box 20, folder 1034, Special Collections of the National Agricultural Library: Edward Fred Knipling Papers, Beltsville, Md.

74. Consolidated DDT Hearings, 547.

75. Ibid., 599–600.

76. Charles F. Wurster, "DDT Proved Neither Essential Nor Safe," (unpublished manuscript) file Pesticide, Scientific Papers, C. Wurster, EDF Archives.

77. Wurster to Mrs. Kurt Woerner, December 11, 1972, file Correspondence Pesticides, EDF Archives.

78. Consolidated DDT Hearings, 2614–15.

79. Ibid., 2638.

80. According to Samuel Rotrosen, there was no direct connection between Borlaug and Montrose. Rotrosen saw an opportunity to publicize that a Nobel Prize winner testified on behalf of DDT, so he arranged a press conference. Starting with the Wisconsin DDT hearing, the EDF held press conferences as well when important witnesses appeared. Samuel Rotrosen (former president of Montrose Chemical Corporation), telephone interview with author, April 3, 2005, New York.

81. Consolidated DDT Hearings, 1337.

82. Ibid., 1343–44.

83. Ibid., 1378.

84. Ibid., 1380.

85. Ibid., 6520.

86. Ibid., 9052.

87. Ibid., 9117.

88. Hearing Examiner's Recommended Findings, Conclusions and Orders, Consolidated DDT Hearings, Environmental Protection Agency, 40 CFR 164.32, April 25, 1972, 95, EDF Archives.

89. Ibid., 95–96.

90. Ibid., 95.

91. "Apologist for DDT," *New York Times*, April 29, 1972.

92. Exceptions of interveners, EDF, et al. to the Hearing Examiner's Recommended Decision and Conduct of Hearing, May 12, 1972, EDF Archives.

93. J. Gordon Edwards, "The Infamous Ruckelshaus DDT Decision," July 24, 1972, *Congressional Record*, S11545–47.

94. See Neier, *Only Judgment*. Recent works on the civil rights movement have made similar arguments. See Klarman, *From Jim Crow to Civil Rights*, and Rosenberg, *The Hollow Hope*.

95. Paul Frost, N.J., to Committee Chairman (Poage), May 18, 1972, Poage Papers, box 110, file 1, BCPM.

96. Donald O. Walter, letter to the editor, *Smithsonian*, December 1970, 6.

97. Opinion of the Administrator, Consolidated DDT Hearings, June 14, 1972, 1, 31, EDF Archives.

98. Ibid., 32.

99. News Release from the office of Congressman George A. Goodling, 19th Pennsylvania District, June 16, 1972, box D 98, Ford Congressional Papers, Gerald R. Ford Presidential Library, Ann Arbor, Mich.

Chapter 7

1. Memo, Fred Ward to W. R. Poage, December 5, 1969, box 89, file 2, W. R. Poage Papers, BCPM.

2. W. R. Poage to Gerald O'Brien, ex-V.P. of American Importers Association, March 25, 1971, box 110, file 3, Poage Papers, BCPM.

3. Bradshaw Mintener to Poage, August 6, 1970, box 89, file 2; Mario Di Pasquale to Poage, August 6, 1970, box 89, file 2; George Boecklin to Poage, box 89, file 2; and Robert H. Smith to Poage, July 7, 1971, box 111, file, Poage Papers, BCPM.

4. See, for example, Soluri, *Banana Cultures*, and Striffler, *In the Shadows of State and Capital*.

5. J. Lyle Littlefield to Poage, November 16, 1970, W. R. Poage Papers, BCPM.

6. Richard E. Lyng to W. R. Poage, October 6, 1970, box 106, file 457, Hyde Murray Papers, BCPM.

7. Bender, *A Nation Among Nations*.

8. Train, *Politics, Pollution, and Pandas*, 121–54.

9. Ibid., 123.

10. The consolidated departments included the Department of Health, Education, and Welfare, the Department of the Interior, the Department of Agriculture, the Atomic Energy Commission, and the Federal Radiation Council.

11. Because the Soviet Union and other Eastern Bloc countries boycotted the conference, much attention was devoted to the environmental consequences of economic and social development along the north-south axis. Björn-Ola Linnér and Henrik Selin, "How It All Began: Global Efforts on Sustainable Development from Stockholm to Rio" (paper presented at 6th Nordic Conference on Environmental Social Sciences, Åbo, Finland, June 12–14, 2003, as part of the panel "Johannesburg: A First Anniversary"). For an earlier summary of the conference, see Rowland, *The Plot to Save the World*.

12. Ward and Dubos, *Only One Earth*, xv.

13. Maurice Strong, introduction to Rowland, *The Plot to Save the World*, viii.

14. Deposition of Samuel Rotrosen, October 13, 1992, *United States of America et al. v. Montrose Chemical Corporation of California et al.*, No. CV 90-3122 AAH (JRX). I would like to thank Terry Kehoe for suggesting this source.

15. DDT: A Review of Scientific and Economic Aspects of the Decision to Ban its Use as a Pesticide, Prepared for Committee on Appropriations, U.S. House of Representatives, Environmental Protection Agency, July 1975, RG 412, Headquarter Records of the Environmental Protection Agency, NACP.

16. Deposition of Samuel Rotrosen, October 13, 1992, *United States of America et al. v. Montrose Chemical Corporation of California et al.*, No. CV 90-3122 AAH (JRX), 409.

17. *Stauffer Chemical Corporation Annual Report*, 1976 and 1979. In 1976, the Stauffer Corporation comprised eight main product lines: industrial chemicals, chemical systems, fertilizer and mining, specialty chemicals, plastics, food ingredients, agricultural chemicals, and international operations.

18. *Stauffer Chemical Corporation Annual Report*, 1979. Sales figures from Stauffer's international sales division consisted of foreign subsidiaries and export. In 1979, export sales totaled $103 million. And for a three-year period ending in 1979, foreign sales increased 2.7 times.

19. Garcia-Johnson, *Exporting Environmentalism*, 86. Dow's foreign sales were 35 percent; Union Carbide's, 26 percent; and Monsanto's, 23 percent.

20. Norris, *Pills, Pesticides and Profits*, 5.

21. General Accounting Office report, *Better Regulation of Pesticide Exports and Pesticide Residues in Imported Food is Essential*, June 22, 1979, 1.

22. Dinham, *The Pesticide Hazard*, 13.

23. Schulberg, "United States Export of Products Banned for Domestic Use," 350, 361.

24. The regulation imposed federal oversight on chemical producers and users. Not only did it establish penalties for misuse, but the legislation also demanded federal control of pesticide use within individual states. *EPA to Ask for Comments on New Pesticides Law*, EPA press release, November 8, 1972, http://www.epa.gov/history/topics/fifra/03.htm.

25. Flippen, "Pests, Pollution, and Politics."

26. "Toxic Legislation," *New York Times*, October 17, 1972.

27. Samuel Rotrosen (former president of Montrose Chemical Corporation), telephone interview with author, April 3, 2005.

28. Barnet and Müller, *Global Reach*, 16.

29. Garcia-Johnson, *Exporting Environmentalism*, 87. Of course, while creating a political body responsive to environmental problems was a significant achievement, enforcement of environmental policies proved to be a difficult task for developing nations, due to lack of financial resources.

30. Shortly thereafter, Ruckelshaus would be named deputy attorney general, a position from which he resigned rather than fire special Watergate prosecutor Archibald Cox, as part of the "Saturday Night Massacre."

31. Train, *Politics, Pollution, and Pandas*, 156.

32. Gladwin Hill, "Environment: Reformers Are Undismayed by the Energy Crisis," *New York Times*, December 2, 1973. This article quotes Texaco chairman Maurice Granville, who claimed that the energy crisis could correct "the overreaction to environmental and consumerist concerns reflected in current statutes and regulations."

33. Train, *Politics, Pollution, and Pandas*, 156.

34. The ruling prohibited Shell Chemical Company, the sole domestic manufacturer, from producing nearly 10 million pounds of the chemical in 1975. See Whitaker, *Striking a Balance*, 136.

35. "Train Stops Manufacture of Heptachlor/Chlordane, Cites Imminent Cancer Risk," EPA press release, July 30, 1975, http://www.epa.gov/history/topics/legal/01.htm. Also see Charles Wurster, "The Case against Aldrin, Dieldrin," *Audubon*, November 1973, 121–24.

36. Richard L. Eldredge, executive director, National Pest Control Association, to John R. Quarles, deputy administrator, EPA, August 6, 1975, box 7, George Humphreys, associate director for environment, 1975–77, Ford Library, Ann Arbor, Mich.

37. Ives, *The Export of Hazard*, 182. In 1985, the nonprofit organization Pesticide Action Network identified DDT, dieldrin, aldrin, heptachlor, chlordane, and seven other chemicals as the "dirty dozen."

38. Daniel P. Resse to Honorable Joseph P. Vigorito, September 15, 1975, box 107, file 483, Hyde Murray Papers; S. D. Gardner to William R. Poage, September 23, 1975, box 107, file 483, Hyde Murray Papers; Erma and Esther Zube, Waco, Tex., to William R. Poage, August 18, 1975, box 1045, file 2, Poage Papers, BCPM.

39. John Baize to Anna Kirk, August 11, 1975, box 107, file 483, Hyde Murray Papers; John C. Baize to Jean Rand, January 27, 1976, box 107, file 483, Hyde Murray Papers, BCPM. Also see Hoberg, *Pluralism by Design*, 150–54.

40. Address by the Honorable Henry A. Kissinger, secretary of state, before the World Food Conference, November 6, 1974, box 10, papers of Paul Leach, Ford Presidential Library, Ann Arbor, Mich.

41. Council of Environmental Quality, annual report, 1975, George Humphreys, box 3/4, Ford Presidential Library, Ann Arbor, Mich.

42. Don Paarlberg, "Changing World Needs for Grains in the 1970's," January 22, 1973, El Batan, Mexico, IX A 1, Special Collections, National Agricultural Library, Beltsville, Md.

43. Council of Environmental Quality, annual report, 1975, box 3, papers of George Humphreys, Ford Presidential Library, Ann Arbor, Mich.

44. John H. Knowles, president, *Rockefeller Foundation Annual Report*, 1975, 1–11.

45. Sandra Postel, *Defusing the Toxics Threat*.

46. Rafael V. Mariano, "The Politics of Pesticides," November 11, 1999, http://www.mindfully.org/Pesticide/Politics-Of-Pesticides-Mariano.htm. Mariano was president of the Peasant Movement of the Philippines (Kilusang Magbubukid ng Pilipinas, or KMP). Like many grassroots movements in the "developing" world, the KMP understood the ecological and human health hazards associated with the global pesticide trade.

47. A split year represents a growing season. See Dinham, *The Pesticide Hazard*, 142.

48. Ibid., 159.

49. General Accounting Office, *Better Regulation of Pesticide Exports and Pesticide Residues in Imported Food is Essential* (Washington, D.C.: Government Printing Office, 1979), 3.

50. Repetto, *Paying the Price*, 1. Repetto conducted a detailed analysis of pesticide subsidies in nine nations and on three continents. Senegal provided the largest subsidy, 89 percent of total cost. Egypt provided 83 percent, and Indonesia provided 82 percent of total cost. At the other end of the spectrum, China and Pakistan subsidized pesticides at 19 percent, while Ecuador provided the next lowest subsidy, 41 percent.

51. Van den Bosch, *The Pesticide Conspiracy*, 30.

52. Alan Riding, "Free Use of Pesticides in Guatemala Takes a Deadly Toll," *New York Times*, November 9, 1977.

53. World Health Organization, *Public Health Impact of Pesticides Used in Agriculture* (Geneva: World Health Organization, 1990), 85–89.

54. Dinham, *The Pesticide Hazard*, 38.

55. Vagliano, "Any Place but There: A Critique of US Hazardous Export Policy," 344.

56. See Bull, *A Growing Problem*, 30–31. Bull reported that in Pakistan malaria rates jumped from 9,500 cases in 1968 to nearly 10 million six years later.

57. Instituto Centroamericano de Investigación y Tecnología (ICAITI), *An Environmental and Economic Study of the Consequences of Pesticide use in Central American Cotton Production, Final Report* (Guatemala: ICAITI, 1976), 3. Also see Williams, *Export Agriculture and the Crisis in Central America*.

58. McCann, *Maize and Grace*, 174–96.

59. Peter Millius and Dan Morgan, "Hazardous Pesticide Sent as Aid," *Washington Post*, December 8, 1976.

60. Bull, *A Growing Problem*, 40.

61. Mark Hosenball, "Karl Marx and the Pajama Game," *Mother Jones*, November 1979, 47.

62. House Committee on Government Operations, *Report on Export of Products Banned By U.S. Regulatory Agencies*, 95th Cong., 2d sess., 1978, H. Doc. 1686, 3, 28.

63. The State Department was not the only domestic institution concerned with the international implications of NEPA. See Burhans, "Exporting NEPA."

64. Christian A. Herter Jr., "Rx for a Cleaner, Healthier World," *Dupont Context*, February 1972, 22.

65. James R. Fowler, special assistant and executive director, A.I.D. Committee on Environmental and Development, to James Frey, Office of Management and Budget, October 22, 1971, RG 286, Records of USAID, HQ of the Administrator, Subject Files, 1971–72, NACP.

66. Charles Warren, "Feeling Uncle Sam's Actions Abroad," *Washington Post*, February 6, 1978.

67. President Carter, Environmental Message to the Congress, May 23, 1977, The Global 2000 Study, RG 429, Records of Organizations in the Executive Office of the President, CEQ, Chairman's Subject Files, 1970–1981, NACP.

68. Russell Train, EPA administrator, Testimony Before the Committee on Foreign Relations, May 5, 1976, box 7, papers of George Humphreys, Ford Library, Ann Arbor, Mich.

69. "Trouble for Export," *Washington Post*, August 27, 1979.

70. Worster, "World without Borders."

71. Sellers, "Body, Place and the State," 31.

72. "Helping the Hand that Feeds Us: Towards an Agriculture for the Long-Term," remarks by the Honorable Russell E. Train, administrator, EPA, before the Board of Trustees Meetings of the Nutrition Foundation, May 2, 1975, Reichley, box 3, Ford Library, Ann Arbor, Mich.

73. Andrew B. Waldo, "A Review of US and International Restrictions on Exports of Hazardous Substances," in Jane H. Ives, *The Export of Hazard*, 20–22; and Newby, "The Export of Pesticides to Developing Nations," 69–82.

74. Stein, *Pivotal Decade*, xi.

75. Executive Order, "Federal Policy Regarding the Export of Banned or Significantly Restricted Materials," *Federal Register* 46 (January 19, 1981), 4659.

76. Joanne Omang, "Carter Limits U.S. Export of Banned Items," *Washington Post*, January 16, 1981.

77. "Unconcern for Hazards is Ascribed to Reagan," *Washington Post*, March 13, 1981.

78. Italy, India, China, and Indonesia continue to manufacture DDT. Recent reports suggest that North Korea also produces the chemical. Only the Indonesian producer P.T. Montrose Pesticido Nusantara has a name similar to that of the former U.S. producer. See Amir Attran and Rajendra Maharaj, "DDT for Malaria Control Should Not Be Banned," *British Medical Journal* 321, no. 7273 (December 2, 2000): 1403–4; and "DDT Factsheet," Pesticide Action Network, UK, http://www.pan-uk.org/pestnews/Actives/ddt.htm.

79. Two suits were filed against Montrose Chemical Corporation of California.

The first resulted from wastewater created at the Montrose Chemical facility and discharged via the municipal sewer system. This runoff produced high concentrations of DDT and other chemicals used in DDT's production in the seabed off Palos Verdes, on the California coast. The other suit was filed by the EPA as a result of high levels of DDT found in and around Montrose's facility in Torrance, California—levels that had threatened or contaminated the groundwater as well. The subsequent cleanup of the facility, which was declared a superfund site by the EPA, followed a series of landmark court rulings that adjudicated the financial responsibilities of interested parties— Montrose Chemical Corporation and Admiral Insurance Company. See *Montrose Chemical Corporation of California v. Admiral Insurance Company*, 10 Cal. 4th 645, S026013. Also see *United States et al. v. Montrose Chemical Corporation of California et al.* For an overview of Montrose Chemical Corporation litigation, see Terence Kehoe and Charles Jacobson, "Environmental Decision Making and DDT Production at Montrose Chemical Corporation of California," *Enterprise and Society* 4, no. 4 (2003): 640–75.

80. James Morton Turner makes a similar argument, but focuses on the politics of the Sagebrush Rebellion and the Wise-Use Movement of the American West. See Turner, "The Specter of Environmentalism," 123–48.

Epilogue

1. *Rockefeller Foundation Annual Report*, 1975, 7.

2. May Berenbaum made a similar argument. See "If Malaria's the Problem, DDT's Not the Only Solution," *Washington Post*, June 5, 2010.

3. Henry I. Miller, "The Uses of DDT," *Wall Street Journal*, August 16, 2007.

4. John Del Signore, "Bedbugs Tormenting Howard Stern, DDT-Hating Hippies Blamed," *The Gothamist*, September 28, 2010, http://gothamist.com/2010/09/28/bedbugs_tormenting_howard_stern_ddt.php.

5. Paul Driessen, "New York's Bedbugs vs. Africa's Malaria Crisis," *CFACT News*, August 21, 2010, http://www.cfact.org/a/1799/New-Yorks-bedbugs-vs-Africas-malaria-crisis.

6. While I have given this idea a lot of thought and have been influenced by the work of Jane Jacobs, Rachel Carson, the British economist E. F. Schumacher, and James C. Scott, all of whom grapple with the question of scale, only a recent lecture by British marketing executive Rory Sutherland provided the language needed to consider these issues. See Rory Sutherland, "Sweat the Small Stuff," TED.com, April 2010, http://www.ted.com/talks/rory_sutherland_sweat_the_small_stuff.html.

7. Sonia Shah, "Learning to Live with Malaria," *New York Times*, October 8, 2010. Also see Shah, *The Fever*.

8. Frederick M. Fishel, "Pesticide Use Trends in the U.S.: Global Comparison," PI-143, Pesticide Information Office, Florida Cooperative Extension Service, Institute of Food and Agricultural Sciences, University of Florida, January 2007. See the EDIS website, http://edis.ifas.ufl.edu.

9. G. M. Calvert et al., "Acute Pesticide Poisoning among Agricultural Workers in the

United States, 1998–2005," *American Journal of Industrial Medicine* 51, no. 12 (December 2008): 883–98.

10. Josef G. Thundiyil et al., "Acute Pesticide Poisoning: A Proposed Classification Tool," *Bulletin of the World Health Organization* 86, no. 3 (2008): 205–9.

11. Of primary concern is DDT's role as an endocrine disruptor, a chemical that interferes with the body's natural endocrine functions, altering normal hormone levels or changing the way hormones travel through the body.

12. "Global Status of DDT and Its Alternatives For Use in Vector Control to Prevent Disease: Background Document for the Preparation of the Business Plan for a Global Partnership to Develop Alternatives to DDT," prepared for the secretariat by Henk van den Berg, UNEP/POPS/DDTBP.1/2, October 2008, 4. North Korea is the only other nation producing DDT, but the chemical is primarily used for domestic purposes.

13. Mason, *The End of the American Century*; Bacevich, *American Empire*; Ferguson, *Colossus*.

14. "Global Status of DDT and Its Alternatives For Use in Vector Control to Prevent Disease: Background Document for the Preparation of the Business Plan for a Global Partnership to Develop Alternatives to DDT," prepared for the secretariat by Henk van den Berg, UNEP/POPS/DDTBP.1/2, October 2008.

Bibliography

Scientific Journals and Trade Magazines

Agricultural Chemicals
American Journal of Hygiene
American Journal of Public Health
and the Nation's Health
American Journal of Tropical Medicine
and Hygiene
Chemical and Engineering News
Chemical Week
Cyanamid Magazine

Journal of Economic Entomology
Journal of the National Malaria Society
Monsanto Magazine
Nature
Organic Gardening and Farming
Science
Science News Letter
Scientific Monthly
Soap and Sanitary Chemicals

Books, Articles, Essays, and Other Sources

Adas, Michael. *Dominance by Design: Technological Imperatives and America's Civilizing Mission*. Cambridge, Mass.: Belknap Press of Harvard University Press, 2006.

Allen, Will. *The War on Bugs*. White River Junction, Vt.: Chelsea Green, 2008.

Anderson, Warwick. *Colonial Pathologies: American Tropical Medicine, Race, and Hygiene in the Philippines*. Durham, N.C.: Duke University Press, 2006.

———. "Going Through the Motions: American Public Health and Colonial 'Mimicry.'" *American Literary History* 14, no. 4 (2002): 686–719.

Augspurger, Michael. *An Economy of Abundant Beauty: Fortune Magazine and Depression America*. Ithaca, N.Y.: Cornell University Press, 2004.

———. "Henry Luce, *Fortune*, and the Attraction of Italian Fascism." *American Studies* 41, no. 1 (Spring 2000): 115–39.

Bacevich, Andrew J. *American Empire: The Realities and Consequences of U.S. Diplomacy*. Cambridge, Mass.: Harvard University Press, 2002.

Bailes, Kendall E., ed. *Environmental History: Critical Issues in Comparative Perspective*. Lanham, Md.: University Press of America, 1985.

Baker, Randolph, Robert W. Herdt, and Beth Rose. *The Rice Economy of Asia*. Washington D.C.: Resources for the Future Press, 1985.

Barnet, Richard J., and Ronald E. Müller. *Global Reach: The Power of the Multinational Corporations*. New York: Simon and Schuster, 1974.

Barsam, Richard Meran. *Nonfiction Film: A Critical History*. Bloomington: Indiana University Press, 1992.

Baumslag, Naomi. *Murderous Medicine: Nazi Doctors, Human Experimentation, and Typhus.* Westport, Conn.: Praeger, 2005.

Baxter, James Phinney, III. *Scientists Against Time.* Boston: Little, Brown, 1946.

Beadle, Christine, and Stephen L. Hoffman. "History of Malaria in the United States Naval Forces at War: World War I through the Vietnam Conflict." *Clinical Infectious Diseases* 16, no. 2 (February 1993): 320–29.

Beatty, Rita Gray. *The DDT Myth: Triumph of the Amateurs.* New York: John Day, 1973.

Becker, Norbert, Marija Zgomba, Dusan Petric, Christine Dahl, Clive Boase, John Lane, and Achim Kaiser. *Mosquitoes and Their Control.* New York: Plenum, 2003.

Bender, Thomas. *A Nation Among Nations: America's Place in World History.* New York: Hill and Wang, 2006.

―――, ed. *Rethinking American History in a Global Age.* Berkeley: University of California Press, 2002.

Bennett, Judith A. "Malaria, Medicine, and Melanesians: Contested Hybrid Spaces in World War II." *Health and History* 8, no. 1 (2006): 27–55.

Berman, Edward H. *The Ideology of Philanthropy: The Influence of the Carnegie, Ford, and Rockefeller Foundations on American Foreign Policy.* Albany: State University of New York Press, 1983.

Berman, Marshall. *All That Is Solid Melts into Air: The Experience of Modernity.* New York: Penguin, 1982.

Bess, Michael. *Light-Green Society: Ecology and Technological Modernity in France, 1960–2000.* Chicago: University of Chicago Press, 2003.

Bickel, Lennard. *Facing Starvation: Norman Borlaug and the Fight Against Hunger.* New York: Reader's Digest Press, 1974.

Birn, Anne-Emanuelle. *Marriage of Convenience: Rockefeller International Health and Revolutionary Mexico.* Vol. 8 of *Rochester Studies in Medical History.* Rochester, N.Y.: University of Rochester Press, 2006.

―――. "Public Health or Public Menace? The Rockefeller Foundation and Public Health in Mexico, 1920–1950." *Voluntas* 7, no. 1 (1996): 35–56.

Borlaug, Norman E. *The Green Revolution, Peace and Humanity.* Mexico: International Maize and Wheat Improvement Center, 1972.

―――. *Mankind and Civilization at Another Crossroad.* Rome: Food and Agriculture Organization of the United Nations, 1971.

Borlaug, Norman E., and International Maize and Wheat Improvement Center. *Norman Borlaug on World Hunger.* San Diego: Bookservice International, 1997.

Borstelmann, Thomas. *The Cold War and the Color Line: American Race Relations in the Global Arena.* Cambridge, Mass.: Harvard University Press, 2003.

Bosso, Christopher J. *Pesticides and Politics.* Pittsburgh: University of Pittsburgh Press, 1987.

Brandt, E. N. *Growth Company: Dow Chemical's First Century.* East Lansing: Michigan State University Press, 1997.

Brar, Karanjot Kaur. *Green Revolution: Ecological Implications*. Delhi: Dominant Publishers and Distributors, 1999.

Briggs, Laura, Gladys McCormick, and J. T. Way. "Transnationalism: A Category of Analysis." *American Quarterly* 60, no. 3 (September 2008): 625–48.

Brinkley, Alan. *The Publisher: Henry Luce and His American Century*. New York: Knopf, 2010.

Brogdon, William G., and Janet C. McAllister. "Insecticide Resistance and Vector Control." *Journal of Agromedicine* 6, no. 2 (1999): 41–58.

Brown, Peter J. "Failure-as-Success: Multiple Meanings of Eradication in the Rockefeller Foundation Sardinia Project, 1946–1951." *Parassitologia* 40, nos. 1 and 2 (June 1998): 117–30.

Bruce-Chwatt, Leonard Jan. "Malaria Eradication at the Crossroads." *Bulletin of the New York Academy of Medicine* 45, no. 10 (October 1969): 999–1012.

———. "Ronald Ross, William Gorgas, and Malaria Eradication." *American Journal of Tropical Medicine and Hygiene* 265 (1977): 1071–79.

Bruce-Chwatt, Leonard Jan, and Julian de Zulueta. *The Rise and Fall of Malaria in Europe: A Historico-Epidemiological Study*. Oxford: Oxford University Press, 1980.

Bruno, Laura A. "The Bequest of the Nuclear Battlefield: Science, Nature, and the Atom During the First Decade of the Cold War." *Historical Studies in the Physical and Biological Sciences* 33, no. 2 (2003): 237–60.

Buell, Frederick. *From Apocalypse to Way of Life: Environmental Crisis in the American Century*. New York: Routledge, 2003.

Buhs, Joshua Blu. "Dead Cows on a Georgia Field: Mapping the Cultural Landscape of the Post–World War II American Pesticide Controversies." *Environmental History* 7, no. 1 (January 2002): 99–121.

———. *The Fire Ant Wars: Nature, Science, and Public Policy in Twentieth-Century America*. Chicago: University of Chicago Press, 2004.

Bull, David. *A Growing Problem: Pesticides and the Third World Poor*. Oxford: Oxfam, 1982.

Bullock, Mary Brown. *An American Transplant: The Rockefeller Foundation and Peking Union Medical College*. Berkeley: University of California Press, 1980.

Burhans, John T. "Exporting NEPA: The Export-Import Bank and the National Environmental Policy Act." *Brooklyn Journal of International Law* 7 (1981): 1–26.

Burnstein, Daniel Eli. *Next to Godliness: Confronting Dirt and Despair in Progressive Era New York City*. Urbana: University of Illinois Press, 2006.

Camp, Roderic. *Entrepreneurs and Politics in Twentieth-Century Mexico*. New York: Oxford University Press, 1989.

Caprotti, Federico. "Malaria and Technological Networks: Medical Geography in the Pontine Marshes, Italy, in the 1930s." *Geographical Journal* 172, no. 2 (June 2006): 145–55.

Caprotti, Federico, and Maria Kaïka. "Producing the Ideal Fascist Landscape: Nature, Materiality and the Cinematic Representation of Land Reclamation in the Pontine Marshes." *Social and Cultural Geography* 9, no. 6 (September 2008): 613–34.

Carson, Rachel. *Silent Spring*. Boston: Houghton Mifflin, 2002.

Carter, Eric D. "'God Bless General Perón': DDT and the Endgame of Malaria Eradication in Argentina in the 1940s." *Journal of the History of Medicine and Allied Sciences* 64, no. 1 (2008): 78–122.

Clarke, Sabine. "A Technocratic Imperial State? The Colonial Office and Scientific Research, 1940–1960." *Twentieth Century British History* 18, no. 4 (January 2007): 453–80.

Cockayne, Emily. *Hubbub: Filth, Noise and Stench in England, 1600–1770*. New Haven, Conn.: Yale University Press, 2007.

Cohen, Lizabeth. *A Consumers' Republic: The Politics of Mass Consumption in Postwar America*. New York: Vintage, 2003.

Coleman, James Samuel, and Rockefeller Foundation. *University Development in the Third World: The Rockefeller Foundation Experience*. Oxford: Pergamon Press, 1993.

Commoner, Barry. *The Closing Circle: Nature, Man, and Technology*. New York: Knopf, 1971.

Connelly, Matthew James. *Fatal Misconception: The Struggle to Control World Population*. Cambridge, Mass.: Belknap Press of Harvard University Press, 2008.

Cooper, Frederick, and Randall Rackard, eds. *International Development and the Social Sciences: Essays on the History and Politics of Knowledge*. Berkeley: University of California Press, 1997.

Corborn, Theo, Dianne Dumanoski, and John Peter Meyers. *Our Stolen Future: Are We Threatening Our Fertility, Intelligence, and Survival? A Scientific Detective Story*. New York: Plume, 1997.

Cotter, Joseph. *Troubled Harvest: Agronomy and Revolution in Mexico, 1880–2002*. Westport, Conn.: Praeger, 2003.

Crichton, Michael. *State of Fear*. New York: Harper Collins, 2004.

Crosby, Alfred W. *Ecological Imperialism: The Biological Expansion of Europe, 900–1900*. 2nd ed. New York: Cambridge University Press, 2004.

Cueto, Marcos. *Cold War, Deadly Fevers: Malaria Eradication in Mexico, 1955–1975*. Washington, D.C.: Woodrow Wilson Center Press, 2007.

———, ed. *Missionaries of Science: The Rockefeller Foundation and Latin America*. Bloomington: Indiana University Press, 1994.

Cullather, Nick. *The Hungry World: America's Cold War Battle against Poverty in Asia*. Cambridge, Mass.: Harvard University Press, 2010.

———. "Miracles of Modernization: The Green Revolution and the Apotheosis of Technology." *Diplomatic History* 28 (April 2004): 227–54.

Culver, John C., and John Hyde. *American Dreamer: The Life and Times of Henry A. Wallace*. New York: W. W. Norton, 2000.

Cushing, Emory. *History of Entomology in World War II*. Washington, D.C.: Smithsonian Institution, 1957.

Dahlberg, Kenneth A. *Beyond the Green Revolution: The Ecology and Politics of Global Agricultural Development*. New York: Plenum Press, 1979.

Daniel, Pete. *Toxic Drift: Pesticides and Health in the Post–World War II South.* Baton Rouge: Louisiana State University Press, 2005.

Datta, Amrita. *Green Revolution in India: Some Unheard Voices.* Kolkata: Best Books, 2006.

Davies, T. G. E., L. M. Field, P. N. R. Usherwood, and M. S. Williamson. "DDT, Pyrethrins, Pyrethroids and Insect Sodium Channels." *IUBMB Life* 59, no. 3 (March 2007): 151–62.

Dean, Warren. *Brazil and the Struggle for Rubber.* New York: Cambridge University Press, 1987.

De Grazia, Victoria. *Irresistible Empire: America's Advance Through Twentieth-Century Europe.* Cambridge, Mass.: Belknap Press of Harvard University Press, 2006.

De Steiguer, Joseph Edward. *The Origins of Modern Environmental Thought.* Tucson: University of Arizona Press, 2006.

Dinham, Barbara, comp. *The Pesticide Hazard: A Global Health and Environmental Audit.* London: Zed Books, 1993.

Doel, Ronald E. "Constituting the Postwar Earth Sciences: The Military's Influence on the Environmental Sciences in the USA after 1945." *Social Studies of Science* 33, no. 5 (October 2003): 635–66.

Doel, Ronald E., and Kristine C. Harper. "Prometheus Unleashed: Science as a Diplomatic Weapon in the Lyndon B. Johnson Administration." *Osiris* 21, no. 1 (January 2006): 66–85.

Dorsey, Kurk, and Mark Lytle. Introduction to *Diplomatic History* 32, no. 4 (2008): 517–18.

Dower, John W. *War without Mercy: Race and Power in the Pacific War.* New York: Pantheon, 1986.

Driessen, Paul. *Eco-Imperialism: Green Power, Black Death.* New Delhi: Academic Foundation, 2006.

Dunlap, Thomas R. *DDT: Scientists, Citizens, and Public Policy.* Princeton, N.J.: Princeton University Press, 1981.

Eckart, W. U., and H. Vondra. "Malaria and World War II: German Malaria Experiments, 1939–45." *Parassitologia* 42, no. 1 (June 2000): 53–58.

Egan, Michael. *Barry Commoner and the Science of Survival: The Remaking of American Environmentalism.* Cambridge, Mass.: MIT Press, 2007.

———. "Forum: Toxic Knowledge: A Mercurial Fugue in Three Parts." *Environmental History* 13, no. 4 (October 2008): 636–42.

Ehrlich, Amy. *Rachel: The Story of Rachel Carson.* San Diego: Harcourt, 2003.

Ellis, John H. *Yellow Fever and Public Health in the New South.* Lexington: University Press of Kentucky, 1992.

Environmental Defense Fund Annual Report, 1971. New York, 1971.

Farley, John. *To Cast Out Disease: A History of the International Health Division of the Rockefeller Foundation, 1913–1951.* Oxford: Oxford University Press, 2004.

Fedunkiw, Marianne. "Malaria Films: Motion Pictures as a Public Health Tool." *American Journal of Public Health* 93, no. 7 (July 2003): 1046–57.

Ferguson, Niall. *Colossus: The Price of America's Empire*. New York: Penguin, 2004.

Fitzgerald, Deborah. "Exporting American Agriculture: The Rockefeller Foundation in Mexico, 1943–53." *Social Studies of Science* 16, no. 3 (1986): 457–83.

Fletcher, W. W. *The Pest War*. New York: Wiley and Sons, 1974.

Flippen, J. Brooks. *Conservative Conservationist: Russell E. Train and the Emergence of American Environmentalism*. Baton Rouge: Louisiana State University Press, 2006.

———. *Nixon and the Environment*. Albuquerque: University of New Mexico Press, 2000.

———. "Pests, Pollution, and Politics: The Nixon Administration's Pesticide Policy." *Agricultural History* 71, no. 4 (Autumn 1997): 442–56.

———. "Richard Nixon, Russell Train, and the Birth of Modern American Environmental Diplomacy." *Diplomatic History* 32, no. 4 (2008): 613–38.

Fogleman, Valerie M. *Guide to the National Environmental Policy Act: Interpretations, Applications, and Compliance*. New York: Quorum Books, 1990.

Fosdick, Raymond B. *The Story of the Rockefeller Foundation*. New York: Harper, 1952.

Fox, Daniel M. "Rockefeller Philanthropy and Modern Biomedicine: International Initiatives from World War I to the Cold War." *Bulletin of the History of Medicine* 78, no. 1 (2004): 248–49.

Gallagher, Nancy Elizabeth. *Egypt's Other Wars: Epidemics and the Politics of Public Health*. Syracuse, N.Y.: Syracuse University Press, 1990.

Garcia-Johnson, Ronie. *Exporting Environmentalism: U.S. Multinational Chemical Corporations in Brazil and Mexico*. Cambridge, Mass.: MIT Press, 2000.

Gaudillière, Jean-Paul. "Science, Technology, and Globalization: Globalization and Regulation in the Biotech World: The Transatlantic Debates over Cancer Genes and Genetically Modified Crops." *Osiris* 21, no. 1 (January 2006): 251–72.

Geissbühler, H. *World Food Production-Environment-Pesticides*. Oxford: Pergamon Press 1979.

Geissler, Erhard, and Jeanne Guillemin. "German Flooding of the Pontine Marshes in World War II." *Politics and the Life Sciences* 29, no. 1 (March 2010): 2–23.

Georghioui, George P. "The Evolution of Resistance." *Annual Review of Ecology and Systematics* 3 (1972): 133–68.

Gilman, Nils. *Mandarins of the Future: Modernization Theory in Cold War America*. Baltimore: Johns Hopkins University Press, 2003.

Goklany, Indur M. *The Precautionary Principle: A Critical Appraisal of Environmental Risk Assessment*. Washington, D.C.: Cato Institute, 2001.

Goldman, Michael. *Imperial Nature: The World Bank and Struggles for Social Justice in the Age of Globalization*. New Haven, Conn.: Yale University Press, 2006.

Goliszek, Andrew. *In the Name of Science: A History of Secret Programs, Medical Research, and Human Experimentation*. New York: St. Martin's Press, 2003.

González, Roberto J. *Zapotec Science: Farming and Food in the Northern Sierra of Oaxaca*. Austin: University of Texas Press, 2001.

Gordin, Michael D. *Five Days in August: How World War II Became a Nuclear War*. Princeton, N.J.: Princeton University Press, 2007.

Gordon, Leonard A. "Wealth Equals Wisdom? The Rockefeller and Ford Foundations in India." *Annals of the American Academy of Political and Social Science* 554 (1997): 104–16.

Gorgas, William Crawford. *Sanitation in Panama.* New York: D. Appleton and Company, 1918.

Gottlieb, Robert. *Forcing the Spring: The Transformation of the American Environmental Movement.* 2nd ed. Washington, D.C.: Island Press, 2005.

Graham, Frank. *Since Silent Spring.* Greenwich, Conn.: Fawcett, 1970.

Graham, Wade. "MexEco? Mexican Attitudes toward the Environment." *Environmental History Review* 15, no. 4 (Winter 1991): 1–17.

Grandin, Greg. *Empire's Workshop: Latin America, the United States, and the Rise of the New Imperialism.* New York: Holt, 2007.

Greene, Benjamin P. *Eisenhower, Science Advice, and the Nuclear Test-Ban Debate, 1945–1963.* Stanford, Calif.: Stanford University Press, 2007.

Grey, Michael R. *New Deal Medicine: The Rural Health Programs of the Farm Security Administration.* Baltimore: Johns Hopkins University Press, 2002.

Grob, Gerald. *The Deadly Truth: A History of Disease in America.* Cambridge, Mass.: Harvard University Press, 2002.

Groshong, Kimm. "The Noisy Response to *Silent Spring*: Placing Rachel Carson's Work in Context." Senior thesis, Pomona College, 2002.

Guber, Deborah Lynn. *The Grassroots of a Green Revolution: Polling America on the Environment.* Cambridge, Mass.: MIT Press, 2003.

Gunter, Valerie J., and Craig K. Harris. "Noisy Winter: The DDT Controversy in the Years Before 'Silent Spring.'" *Rural Sociology* 63, no. 2 (1998): 179–98.

Hanhimäki, Jussi M. "Global Visions and Parochial Politics: The Persistent Dilemma of the 'American Century.'" *Diplomatic History* 27, no. 4 (September 2003): 423–47.

Hardenburg, Walter Ernest. *Mosquito Eradication.* New York: McGraw-Hill, 1922.

———. *The Putumayo—The Devil's Paradise—Travels in the Peruvian Amazon Region and an Account of the Atrocities Committed upon the Indians Therein.* London: T. Fisher Unwin, 1912.

Harmond, Richard P. "Robert Cushman Murphy and Environmental Issues on Long Island." *Long Island Historical Journal* 8, no. 1 (1995): 76–82.

Harte, John, and Robert H. Socolow, eds. *Patient Earth.* New York: Holt, Rinehart and Winston, 1971.

Hays, C. W. "The United States Army and Malaria Control in World War II." *Parassitologia* 42 (2000): 47–52.

Hays, J. N. *The Burdens of Disease: Epidemics and Human Response in Western History.* Rev. ed. New Brunswick, N.J.: Rutgers University Press, 2010.

Hays, Samuel P. *A History of Environmental Politics since 1945.* Pittsburgh: University of Pittsburgh Press, 2000.

———. "The Structure of Environmental Politics since World War II." *Journal of Social History* 14, no. 4 (Summer 1981): 719–38.

Hazell, P. B. R. *The Green Revolution Reconsidered: The Impact of High-Yielding Rice Varieties in South India.* Baltimore: Johns Hopkins University Press, 1991.

Hazlett, Maril. "The Story of *Silent Spring* and the Ecological Turn." Ph.D. diss., University of Kansas, 2003.

———. "'Woman vs. Man vs. Bugs': Gender and Popular Ecology in Early Reactions to *Silent Spring*." *Environmental History* 9, no. 4 (2004): 701–29.

Heffelfinger, Elizabeth. "Foreign Policy, Domestic Fiction: Government-Sponsored Documentaries and Network Television Promote the Marshall Plan at Home." *Historical Journal of Film, Radio and Television* 28, no. 1 (2008): 1–21.

Henkin, Harmon, Martin Merta, and James Staples. *The Environment, the Establishment, and the Law.* Boston: Houghton Mifflin, 1971.

Hersey, John. *Hiroshima.* New York: A. A. Knopf, 1985.

Hershfield, Joanne. *Imagining La Chica Moderna: Women, Nation, and Visual Culture in Mexico, 1917–1936.* Durham, N.C.: Duke University Press, 2008.

Herzstein, Robert E. *Henry R. Luce, Time, and the American Crusade in Asia.* New York: Cambridge University Press, 2006.

Hewitt de Alcantara, Cynthia. *Modernizing Mexican Agriculture: Socioeconomic Implications of Technological Change, 1940–1970.* Geneva: United Nations Research Institute for Social Development, 1976.

Hoberg, George. *Pluralism by Design: Environmental Policy and the American Regulatory State.* New York: Praeger, 1992.

Hodge, Joseph Morgan. *Triumph of the Expert: Agrarian Doctrines of Development and the Legacies of British Colonialism.* Athens: Ohio University Press, 2007.

Hogan, Michael J., ed. *The Ambiguous Legacy: U.S. Foreign Relations in the "American Century."* New York: Cambridge University Press, 1999.

Holloway, Leo D. *The DDT System of Place-Value Notation.* New York: Vantage Press, 1976.

Horne, Gerald. *Race War: White Supremacy and the Japanese Attack on the British Empire.* New York: New York University Press, 2004.

Hossain, Mahabub. *Nature and Impact of the Green Revolution in Bangladesh.* Washington, D.C.: International Food Policy Research Institute, 1988.

Howard, L. O. *A History of Applied Entomology: (Somewhat Anecdotal).* Washington, D.C.: Smithsonian Institution, 1930.

———. *Mosquitoes: How They Live; How They Carry Disease; How They Are Classified; How They May Be Destroyed.* New York: McClure, Phillips and Co., 1902.

Howe, Peter. *Shooting Under Fire: The World of the War Photographer.* New York: Artisan, 2002.

Hoy, Marjorie A. "Myths, Models and Mitigation of Resistance to Pesticides." *Philosophical Transactions of the Royal Society B: Biological Science* 353 (1998): 1787–95.

Hughes, Thomas. *Human-Built World: How to Think About Technology and Culture.* Chicago: University of Chicago Press, 2004.

Humphreys, Margaret. "Kicking a Dying Dog: DDT and the Demise of Malaria in the American South, 1942–1950." *Isis* 87, no. 1 (March 1996): 1–17.

————. *Malaria: Poverty, Race, and Public Health in the United States*. Baltimore: Johns Hopkins University Press, 2001.

————. *Yellow Fever and the South*. Baltimore: Johns Hopkins University Press, 1999.

Hunt, Michael. *Ideology and U.S. Foreign Policy*. New Haven, Conn.: Yale University Press, 1988.

Hurt, R. Douglas. *American Agriculture: A Brief History*. Rev. ed. West Lafayette, Ind.: Purdue University Press, 2002.

Hynes, H. Patricia. *The Recurring Silent Spring*. New York: Pergamon Press, 1989.

Iriye, Akira. "Environmental History and International History." *Diplomatic History* 32, no. 4 (2008): 643–46.

————. *Global Community: The Role of International Organizations in the Making of the Contemporary World*. Berkeley: University of California Press, 2002.

Ives, Jane H., ed. *The Export of Hazard: Transnational Corporations and Environmental Control Issues*. Boston: Routledge, 1985.

Jackson, J. "Cognition and Global Malaria Eradication Programme." *Parassitologia* 40, nos. 1 and 2 (June 1998): 193–216.

Jacobs, Wilbur R. "The Great Despoliation: Environmental Themes in American Frontier History." *Pacific Historical Review* 47, no. 1 (February 1978): 1–26.

Jasanoff, Sheila. "Biotechnology and Empire: The Global Power of Seeds and Science." *Osiris* 21, no. 1 (January 2006): 273–92.

Jennings, Bruce H. *Foundations of International Agricultural Research: Science and Politics in Mexican Agriculture*. Boulder, Colo.: Westview Press, 1988.

Joseph, Gilbert M., Anne Rubenstein, and Eric Zolov, eds. *Fragments of a Golden Age: The Politics of Culture in Mexico since 1940*. Durham, N.C.: Duke University Press, 2001.

Joy, Robert J. "Malaria in American Troops in the South and Southwest Pacific in World War II." *Medical History* 43, no. 2 (April 1999): 192–207.

Judt, Tony. *Postwar: A History of Europe Since 1945*. New York: Penguin, 2006.

King, Nicholas B. "The Influence of Anxiety: September 11th, Bioterrorism, and American Public Health." *Journal of the History of Medicine and the Allied Sciences* 58, no. 4 (2003): 433–41.

————. "Security, Disease, Commerce: Ideologies of Post-Colonial Global Health." *Social Studies of Science* 32, nos. 5 and 6 (2002): 763–89.

Kinkela, David. "The Ecological Landscapes of Jane Jacobs and Rachel Carson." *American Quarterly* 61, no. 4 (December 2009): 905–28.

Klarman, Michael J. *From Jim Crow to Civil Rights: The Supreme Court and the Struggle for Racial Equality*. New York: Oxford University Press, 2004.

Kleinman, Mark L. *A World of Hope, a World of Fear: Henry A. Wallace, Reinhold Niebuhr, and American Liberalism*. Columbus: Ohio State University Press, 2000.

Klingle, Matthew W. "Spaces of Consumption in Environmental History." *History and Theory* 42, no. 4 (December 2003): 94–110.

Knipling, E. F. "Use of Insects for Their Own Destruction." *Journal of Economic Entomology* 53 (June 1960): 415–20.

Kozol, Wendy. *Life's America: Family and Nation in Postwar Photojournalism.* Philadelphia: Temple University Press, 1994.

Krige, John. "Atoms for Peace, Scientific Internationalism, and Scientific Intelligence." *Osiris* 21, no. 1 (January 2006): 161–81.

Kruse, Kevin M. *White Flight: Atlanta and the Making of Modern Conservatism.* Princeton, N.J.: Princeton University Press, 2007.

Langston, Nancy. *Toxic Bodies: Hormone Disruptors and the Legacy of DES.* New Haven, Conn.: Yale University Press, 2010.

Lassiter, Matthew. *The Silent Majority: Suburban Politics in the Sunbelt South.* Princeton, N.J.: Princeton University Press, 2007.

Latham, Michael. *Modernization as Ideology: American Social Science and "Nation Building" in the Kennedy Era.* Chapel Hill: University of North Carolina Press, 2000.

Latour, Bruno. *Pandora's Hope: Essays on the Reality of Science Studies.* Cambridge, Mass.: Harvard University Press, 1999.

———. *Science in Action: How to Follow Scientists and Engineers Through Society.* Cambridge, Mass.: Harvard University Press, 1987.

Lear, Linda. "Bombshell in Beltsville: The USDA and the Challenge of 'Silent Spring.'" *Agricultural History* 66, no. 2 (Spring, 1992): 151–70.

———. *Rachel Carson: Witness for Nature.* New York: Henry Holt, 1997.

Leary, James Cornelius. *DDT and the Insect Problem.* New York: McGraw-Hill, 1946.

Leavitt, Judith W., and Ronald L. Numbers. *Sickness and Health in America: Readings in the History of Medicine and Public Health.* Rev. ed. Madison: University of Wisconsin Press, 1997.

Lekan, Thomas. "Globalizing American Environmental History." *Environmental History* 10, no. 1 (January 2005): 7 parts.

Lenoir, Timothy, ed. *Inscribing Science: Scientific Texts and the Materiality of Communication.* Stanford, Calif.: Stanford University Press, 1998.

Le Prince, Joseph A. *Mosquito Control in Panama: The Eradication of Malaria and Yellow Fever in Cuba and Panama.* New York: Putnam, 1916.

Leslie, Stuart W., and Robert Kargon. "Exporting MIT: Science, Technology, and Nation-Building in India and Iran." *Osiris* 21, no. 1 (January 2006): 110–30.

Levy, Barry S., and Victor W. Sidel, eds. *War and Public Health.* New York: Oxford University Press, 1997.

Logan, John. *Ente regionale per la lotta anti-anofelica in Sardegna [The Sardinian Project: An Experiment in the Eradication of an Indigenous Malarious Vector].* Baltimore: Johns Hopkins University Press, 1953.

Luce, Henry. "The American Century." *Life*, February 17, 1941, 61–65.

Lundestad, Geir. "'Empire by Invitation' in the American Century." *Diplomatic History* 23, no. 2 (1999): 189–217.

———. *The United States and Western Europe since 1945: From "Empire" by Invitation to Transatlantic Drift.* New York: Oxford University Press, 2003.

Lutts, Ralph H. "Chemical Fallout: Rachel Carson's *Silent Spring*, Radioactive Fallout, and the Environmental Movement." *Environmental Review* 9 (1985): 214–25.

Lytle, Mark H. *The Gentle Subversive: Rachel Carson, Silent Spring, and the Rise of the Environmental Movement*. New York: Oxford University Press, 2007.

Maher, Neil M. *Nature's New Deal: The Civilian Conservation Corps and the Roots of the American Environmental Movement*. New York: Oxford University Press, 2007.

Mallet, James. "The Evolution of Insecticide Resistance: Have the Insects Won?" *Trends in Ecology and Evolution* 4, no. 11 (November 1989): 336–40.

Manela, Erez. "A Pox on Your Narrative: Writing Disease Control into Cold War History." *Diplomatic History* 34, no. 2 (April 2010): 299–323.

Marco, Gino J., Robert M. Hollingworth, and William Durham, eds. *Silent Spring Revisited*. Washington, D.C.: American Chemical Society, 1987.

Mariano, Marco, ed. *Defining the Atlantic Community: Culture, Intellectuals, and Policies in the Mid-Twentieth Century*. New York: Routledge, 2010.

Mason, David S. *The End of the American Century*. Boston: Rowman and Littlefield, 2008.

May, Elaine Tyler. *Homeward Bound: American Families in the Cold War Era*. New York: Basic Books, 1988.

McAlister, Melani. *Epic Encounters: Culture, Media, and U.S. Interests in the Middle East, 1945–2000*. Berkeley: University of California Press, 2001.

McBride, David. *Missions for Science: U.S. Technology and Medicine in America's African World*. New Brunswick, N.J.: Rutgers University Press, 2002.

McCann, James C. *Maize and Grace: Africa's Encounter with a New World Crop, 1500–2000*. Cambridge, Mass.: Harvard University Press, 2005.

McCook, Stuart. *States of Nature: Science, Agriculture, and the Environment in the Spanish Caribbean, 1760–1940*. Austin: University of Texas Press, 2002.

McGrath, Patrick, J. *Scientists, Business, and the State, 1890–1960*. Chapel Hill: University of North Carolina Press, 2001.

McMillen, Wheeler. *Bugs or People?* New York: Appleton-Century, 1965.

McNeill, J. R. *Mosquito Empires: Ecology and War in the Greater Caribbean, 1620–1914*. New York: Cambridge University Press, 2010.

———. *Something New under the Sun: An Environmental History of the Twentieth-Century World*. New York: W. W. Norton, 2000.

McNeill, J. R., and Corinna R. Unger, eds. *Environmental Histories of the Cold War*. New York: Cambridge University Press, 2010.

McWilliams, James E. *American Pests: The Losing War on Insects from Colonial Times to DDT*. New York: Columbia University Press, 2008.

Merchant, Carolyn. *American Environmental History: An Introduction*. New York: Columbia University Press, 2007.

Merrill, Dennis. *Bread and the Ballot: The United States and India's Economic Development, 1947–1963*. Chapel Hill: University of North Carolina Press, 1990.

Miller, Clark A. "'An Effective Instrument of Peace': Scientific Cooperation as an Instrument of U.S. Foreign Policy, 1938–1950." *Osiris* 21, no. 1 (January 1, 2006): 133–60.

Miller, Daniel R. "Marriage of Convenience: Rockefeller International Health and Revolutionary Mexico." *Americas* 65, no. 2 (2008): 272–74.

Miller, Shawn William. *An Environmental History of Latin America*. New York: Cambridge University Press, 2007.

Misa, Thomas J., Philip Brey, and Andrew Feenberg, eds. *Modernity and Technology*. Cambridge, Mass.: MIT Press, 2004.

Mitchell, Angus, ed. *The Amazon Journal of Roger Casement*. Dublin: Anaconda Editions, 2000.

Mitchell, Timothy. *Rule of Experts: Egypt, Techno-Politics, Modernity*. Berkeley: University of California Press, 2002.

Moeller, Susan D. *Shooting War: Photography and the American Experience of Combat*. New York: Basic Books, 1989.

Molina, Natalia. *Fit to Be Citizens? Public Health and Race in Los Angeles, 1879–1939*. Berkeley: University of California Press, 2006.

Moreno, Julio. *Yankee Don't Go Home! Mexican Nationalism, American Business Culture, and the Shaping of Modern Mexico, 1920–1950*. Chapel Hill: University of North Carolina Press, 2003.

Morris, John G. *Get the Picture: A Personal History of Photojournalism*. Chicago: University of Chicago Press, 2004.

Mrázek, Rudolf. *Engineers of Happy Land: Technology and Nationalism in a Colony*. Princeton, N.J.: Princeton University Press, 2002.

Muldavin, Joshua. "The Paradoxes of Environmental Policy and Resource Management in Reform-Era China." *Economic Geography* 76, no. 3 (July 2000): 244–71.

Mumford, Lewis. *Technics and Civilization*. New York: Harcourt, Brace, 1934.

Murphy, Michelle. "Forum: Chemical Regimes of Living." *Environmental History* 13, no. 4 (October 2008): 695–704.

Murphy, Priscilla Coit. *What a Book Can Do: The Publication and Reception of Silent Spring*. Amherst: University of Massachusetts Press, 2005.

Murray, Douglas L. *Cultivating Crisis: The Human Cost of Pesticides in Latin America*. Austin: University of Texas Press, 1994.

Myren, Delbert T. "The Rockefeller Foundation Program in Corn and Wheat in Mexico." In *Subsistence Agriculture and Economic Development*, edited by Clifton R. Wharton. Chicago: Aldine, 1969.

Nash, Linda. "Forum: Purity and Danger: Historical Reflections on the Regulation of Environmental Pollutants." *Environmental History* 13, no. 4 (October 2008): 651–58.

———. "The Fruits of Ill-Health: Pesticides and Workers' Bodies in Post–World War II California." *Osiris* 19, 2nd series (2004): 203–19.

———. *Inescapable Ecologies: A History of Environment, Disease, and Knowledge*. Berkeley: University of California Press, 2007.

Nash, Roderick. *Wilderness and the American Mind*. 3rd ed. New Haven, Conn.: Yale University Press, 1982.

Needham, Cynthia. *Global Disease Eradication: The Race for the Last Child*. Washington, D.C.: ASM Press, 2003.

Neier, Aryeh. *Only Judgment: The Limits of Litigation in Social Change*. Middletown, Conn.: Wesleyan University Press, 1982.

Newby, Maureen Ellen. "The Export of Pesticides to Developing Nations: Implications for the Structure of International Interaction." Ph.D. diss., University of Oregon, 1991.

Norris, Ruth, ed. *Pills, Pesticides and Profits.* Croton-on-Hudson, N.Y.: North River Press, 1982.

Norwood, Vera L. "Heroines of Nature: Four Women Respond to the American Landscape." *Environmental Review* 8, no. 1 (Spring 1984): 34–56.

Oreskes, Naomi, and Erik M. Conway. *Merchants of Doubt: How a Handful of Scientists Obscured the Truth on Issues from Tobacco Smoke to Global Warming.* London: Bloomsbury Press, 2010.

Osborne, Fairfield. *Our Plundered Planet.* Boston: Little, Brown, 1948.

Oshinsky, David M. *Polio: An American Story.* New York: Oxford University Press, 2005.

Paarlberg, Don. *Norman Borlaug, Hunger Fighter.* Washington, D.C.: Foreign Economic Development Service, U.S. Department of Agriculture, 1970.

Paarlberg, Robert L. *Starved for Science: How Biotechnology Is Being Kept Out of Africa.* Cambridge, Mass.: Harvard University Press, 2008.

Packard, Randall M. *The Making of a Tropical Disease: A Short History of Malaria.* Baltimore: Johns Hopkins University Press, 2007.

Packard, Randall M., and Peter J. Brown. "Malaria, *Miseria*, and Underpopulation in Sardinia: The 'Malaria Block Development' Cultural Model." *Medical Anthropology* 17, no. 3 (1997): 239–54.

Packard, Randall M., and P. Gadelha. "A Land Filled with Mosquitoes: Fred L. Soper, the Rockefeller Foundation, and the *Anopheles Gambiae* Invasion of Brazil." *Parassitologia* 36 (1994): 197–213.

Pampana, Emilio. *A Textbook of Malaria Eradication.* 2nd ed. London: Oxford University Press, 1969.

Pampana, Emilio, and the World Health Organization. *Malaria: A World Problem.* Geneva: World Health Organization, 1955.

Patterson, Gordon. *The Mosquito Crusades: A History of the American Anti-Mosquito Movement from the Reed Commission to the First Earth Day.* New Brunswick, N.J.: Rutgers University Press, 2009.

———. *The Mosquito Wars: A History of Mosquito Control in Florida.* Gainesville: University Press of Florida, 2004.

Pearce, Kimber Charles. *Rostow, Kennedy, and the Rhetoric of Foreign Aid.* East Lansing: Michigan State University Press, 2001.

Perlstein, Rick. *Nixonland: The Rise of a President and the Fracturing of America.* New York: Scribner, 2008.

Perkins, John H. *Geopolitics and the Green Revolution: Wheat, Genes, and the Cold War.* New York: Oxford University Press, 1997.

———. *Insects, Experts, and the Insecticide Crisis: The Quest for New Pest Management Strategies.* New York: Plenum Press, 1982.

———. "Insects, Food, and Hunger: The Paradox of Plenty for U.S. Entomology, 1920–1970." *Environmental Review* 7, no. 1 (Spring 1983): 71–96.

———. "Reshaping Technology in Wartime: The Effect of Military Goals on Entomological Research and Insect-Control Practices." *Technology and Culture* 19, no. 2 (April 1978): 169–86.

Perkovich, George. *India's Nuclear Bomb: The Impact on Global Proliferation.* Berkeley: University of California Press, 1999.

Peterson, Peter A., and Angelo Bianchi. *Maize Genetics and Breeding in the 20th Century.* New York: World Scientific, 1998.

Pierce, John R. *Yellow Jack: How Yellow Fever Ravaged America and Walter Reed Discovered Its Deadly Secrets.* Hoboken, N.J.: Wiley, 2005.

Pimentel, David, and Hugh Lehman, eds. *The Pesticide Question: Environment, Economics, and Ethics.* New York: Chapman Hill, 1993.

Poleman, Thomas T. *The Papaloapan Project: Agricultural Development in the Mexican Tropics.* Stanford, Calif.: Stanford University Press, 1964.

Poole, Deborah. *Vision, Race, and Modernity: A Visual Economy of the Andean Image World.* Princeton, N.J.: Princeton University Press, 1997.

Posgate, W. D. "Fertilizers for India's Green Revolution: The Shaping of Government Policy." *Asian Survey* 4, no. 8 (August 1974): 733–50.

Postel, Sandra. *Defusing the Toxics Threat: Controlling Pesticides and Industrial Waste.* Washington, D.C.: Worldwatch, 1987.

Prashad, Vinjay. *The Darker Nations: A People's History of the Third World.* New York: New Press, 2008.

Price-Smith, Andrew T. *Contagion and Chaos: Disease, Ecology, and National Security in the Era of Globalization.* Cambridge, Mass.: MIT Press, 2009.

———. *The Health of Nations: Infectious Disease, Environmental Change, and Their Effects on National Security and Development.* Cambridge, Mass.: MIT Press, 2001.

Proctor, Robert. *Racial Hygiene: Medicine Under the Nazis.* Cambridge, Mass.: Harvard University Press, 1988.

Rauchway, Eric. *Blessed Among Nations: How the World Made America.* New York: Hill and Wang, 2006.

Ray, Dixy Lee. *Environmental Overkill: Whatever Happened to Common Sense?* With Lou Guzzo. New York: Harper Perennial, 1993.

Reagan, Leslie J., Nancy Tomes, and Paula A. Treichler, eds. *Medicine's Moving Pictures: Medicine, Health, and Bodies in American Film and Television.* Rochester, N.Y.: University of Rochester Press, 2007.

Repetto, Robert. *Paying the Price: Pesticide Subsidies in Developing Countries.* Washington, D.C.: World Resource Institute, 1985.

Revoldt, Daryl. *Raymond B. Fosdick: Reform, Internationalism, and the Rockefeller Foundation.* Akron, Ohio: University of Akron, 1982.

Rist, Gilbert. *The History of Development: From Western Origins to Global Faith.* London: Zed Books, 2002.

Rivas, Darlene. *Missionary Capitalist: Nelson Rockefeller in Venezuela.* Chapel Hill: University of North Carolina Press, 2001.

Robbins, Paul. *Lawn People: How Grasses, Weeds, and Chemicals Make Us Who We Are.* Philadelphia: Temple University Press, 2007.

Roberts, Donald, and Richard Tren. *The Excellent Powder: DDT's Political and Scientific History.* Indianapolis: Dog Ear Publishing, 2010.

Roberts, Jody A., and Nancy Langston. "Toxic Bodies/Toxic Environments: An Interdisciplinary Forum." *Environmental History* 13, no. 4 (October 2008): 629–35.

Robertson, Thomas. "The Population Bomb: Population Growth, Globalization, and American Environmentalism, 1945–1980." Ph.D. diss., University of Wisconsin, Madison, 2005.

———. "'This Is the American Earth': American Empire, the Cold War, and American Environmentalism." *Diplomatic History* 32, no. 4 (2008): 561–84.

Rockefeller Foundation Annual Report, 1944–55, 1962, and 1975. New York: Corporate Publishing.

Roeder, George H. *The Censored War: American Visual Experience During World War II.* New Haven, Conn.: Yale University Press, 1993.

Rome, Adam. *The Bulldozer in the Countryside: Suburban Sprawl and the Rise of American Environmentalism.* New York: Cambridge University Press, 2001.

———. "'Give Earth a Chance': The Environmental Movement and the Sixties." *Journal of American History* 90, no. 2 (September 2003): 525–54.

Rosenberg, Charles E. *The Cholera Years: The United States in 1832, 1849, and 1866.* Chicago: University of Chicago Press, 1987.

Rosenberg, Emily S. *Financial Missionaries to the World: The Politics and Culture of Dollar Diplomacy, 1900–1930.* Cambridge, Mass.: Harvard University Press, 1999.

———. *Spreading the American Dream: American Economic and Cultural Expansion, 1890–1945.* New York: Hill and Wang, 1982.

Rosenberg, Gerald N. *The Hollow Hope: Can Courts Bring About Social Change?* 2nd ed. Chicago: University of Chicago Press, 2008.

Rostow, W. W. *Eisenhower, Kennedy, and Foreign Aid.* Austin: University of Texas Press, 1985.

Rotter, Andrew J. *Comrades at Odds: The United States and India, 1947–1992.* Ithaca, N.Y.: Cornell University Press, 2000.

———. *Hiroshima: The World's Bomb.* Oxford: Oxford University Press, 2008.

Rowland, Wade. *The Plot to Save the World: The Life and Times of the Stockholm Conference on the Human Environment.* Toronto: Clarke Irwin, 1973.

Russell, Edmund. "The Strange Career of DDT: Experts, Federal Capacity, and Environmentalism in World War II." *Technology and Culture* 40, no. 4 (1999): 770–96.

———. *War and Nature: Fighting Humans and Insects with Chemicals from World War I to Silent Spring.* Cambridge: Cambridge University Press, 2001.

Russell, Paul. Introduction to *Preventative Medicine in World War II,* vol. 6. Washington, D.C.: Office of the Surgeon General, 1963.

———. *Man's Mastery of Malaria.* London: Oxford University Press, 1955.

Sale, Kirkpatrick. *The Green Revolution: The American Environmental Movement, 1962–1992.* New York: Hill and Wang, 1993.

Sallares, Robert. *Malaria and Rome: A History of Malaria in Ancient Italy.* New York: Oxford University Press, 2002.

Sanderson, Steven E. *The Transformation of Mexican Agriculture: International Structure and the Politics of Rural Change.* Princeton, N.J.: Princeton University Press, 1986.

Scheffer, Victor B. *The Shaping of Environmentalism in America.* Seattle: University of Washington Press, 1991.

Schneider, Joanne. *The Age of Romanticism.* Westport, Conn.: Greenwood Press, 2007.

Schulberg, Francine. "United States Export of Products Banned for Domestic Use." *Harvard International Law Journal* 20, no. 2 (Spring 1979): 331–81.

Schuler, Friedrich E. *Mexico between Hitler and Roosevelt: Mexican Foreign Relations in the Age of Lázaro Cárdenas, 1934–1940.* Albuquerque: University of New Mexico Press, 1999.

Schulman, Bruce. *The Seventies: The Great Shift in American Culture, Society, and Politics.* Cambridge, Mass.: Da Capo, 2002.

Schulman, Bruce J., and Julian E. Zelizer, eds. *Rightward Bound: Making America Conservative in the 1970s.* Cambridge, Mass.: Harvard University Press, 2008.

Scott, James. *Seeing Like a State: How Certain Schemes to Improve the Human Condition Have Failed.* New Haven, Conn.: Yale University Press, 1998.

Seigel, Micol. "Beyond Compare: Comparative Method after the Transnational Turn." *Radical History Review* 91 (Winter 2005): 62–90.

Sellers, Christopher. "Body, Place and the State: The Makings of an 'Environmentalist' Imaginary in the Post–World War II U.S." *Radical History Review* 74 (Spring 1999): 31–64.

Sellers, Christopher, Gregg Mitman, and Michelle Murphy. "A Cloud over History." *Osiris* 19 (2005): 1–17.

Sexton, Jay. "The Global View of the United States." *Historical Journal* 48, no. 1 (March 2005): 261–76.

Shabecoff, Philip. *Poisoned Profits: The Toxic Assault on Our Children.* New York: Random House, 2008.

Shah, Nayan. *Contagious Divides: Epidemics and Race in San Francisco's Chinatown.* Berkeley: University of California Press, 2001.

Shah, Sonia. *The Fever: How Malaria Has Ruled Humankind for 500,000 Years.* New York: Farrar, Straus and Giroux, 2010.

Shaplen, Robert. *Toward the Well-Being of Mankind: Fifty Years of the Rockefeller Foundation.* Garden City, N.Y.: Doubleday, 1964.

Shiva, Vandana. *The Violence of the Green Revolution: Third World Agriculture, Ecology, and Politics.* London: Zed Books, 1991.

Sideris, Lisa H., and Kathleen Dean Moore, eds. *Rachel Carson: Legacy and Challenge.* Albany: State University of New York Press, 2008.

Simmons, I. G. *Global Environmental History.* Chicago: University of Chicago Press, 2008.

Simonian, Lane. *Defending the Land of the Jaguar: A History of Conservation in Mexico.* Austin: University of Texas Press, 1995.

Slater, Leo B. *War and Disease: Biomedical Research on Malaria in the Twentieth Century.* New Brunswick, N.J.: Rutgers University Press, 2009.

Slater, Leo Barney, and Margaret Humphreys. "Parasites and Progress: Ethical Decision-Making and the Santee-Cooper Malaria Study, 1944–1949." *Perspectives in Biology and Medicine* 51, no. 1 (2007): 103–20.

Smith, Anthony. *Explorers of the Amazon.* Chicago: University of Chicago Press, 1990.

Smith, Susan Lynn. "Mustard Gas and American Race-Based Human Experimentation in World War II." *Journal of Law, Medicine and Ethics* 36, no. 3 (2008): 517–21.

———. *Sick and Tired of Being Sick and Tired: Black Women's Health Activism in America, 1890–1950.* Philadelphia: University of Pennsylvania Press, 1995.

Smith, Tony. *America's Mission: The United States and the Worldwide Struggle for Democracy in the Twentieth Century.* Princeton, N.J.: Princeton University Press, 1994.

Snowden, Frank Martin. *The Conquest of Malaria: Italy, 1900–1962.* New Haven, Conn.: Yale University Press, 2006.

Soluri, John. *Banana Cultures: Agriculture, Consumption, and Environmental Change in Honduras and the United States.* Austin: University of Texas Press, 2005.

Sonnenfeld, David A. "Mexico's 'Green Revolution,' 1940–1980: Towards an Environmental History." *Environmental History Review* 16, no. 4 (Winter 1992): 29–52.

Sontag, Susan. *Regarding the Pain of Others.* New York: Farrar, Straus and Giroux, 2003.

Soper, Fred Lowe. *Building the Health Bridge: Selections from the Works of Fred L. Soper.* Bloomington: Indiana University Press, 1970.

———. *Ventures in World Health: The Memoirs of Fred Lowe Soper.* Edited by John Duffy. Washington, D.C.: Pan American Health Organization, Pan American Sanitary Bureau, Regional Office of the World Health Organization, 1977.

Spenser, Daniela. *The Impossible Triangle: Mexico, Soviet Russia, and the United States in the 1920s.* Durham, N.C.: Duke University Press, 1999.

Stakman, E. C. "Reminiscences of Elvin Charles Stakman: Oral History, 1970." Interview by Pauline Madow. Oral History Research Office, Columbia University.

Stakman, E. C., Richard Bradfield, and Paul C. Mangelsdorf. *Campaigns Against Hunger.* Cambridge, Mass.: Belknap Press of Harvard University Press, 1967.

Staples, Amy L. S. *The Birth of Development: How the World Bank, Food and Agriculture Organization, and World Health Organization Changed the World, 1945–1965.* Kent, Ohio: Kent State University Press, 2006.

Stapleton, Darwin H. "The Dawn of DDT and Its Experimental Use by the Rockefeller Foundation in Mexico, 1943–1952." *Parassitologia* 40, no. 2 (June 1998): 149–58.

———. "Internationalism and Nationalism: the Rockefeller Foundation, Public Health, and Malaria in Italy, 1923–1951." *Parassitologia* 42 (June 2000): 127–34.

———. "Lessons of History? Anti-Malaria Strategies of the International Health Board and the Rockefeller Foundation from the 1920s to the Era of DDT." *Public Health Reports* 119, no. 2 (n.d.): 206–15.

———. "A Lost Chapter in the Early History of DDT: The Development of Anti-

Typhus Technologies by the Rockefeller Foundation's Louse Laboratory, 1942–
1944." *Technology and Culture* 46, no. 3 (2005): 513–40.

———. "The Short-Lived Miracle of DDT." *American Heritage of Invention and Technology* 15, no. 3 (2000): 34–41.

Stauffer Chemical Corporation Annual Report, 1976 and 1979. Westport, Conn.: Stauffer Chemical Corp.

Stein, Judith. *The Pivotal Decade: How the United States Traded Factories for Finance in the Seventies.* New Haven, Conn.: Yale University Press, 2010.

Steinberg, Theodore. *American Green: The Obsessive Quest for the Perfect Lawn.* New York: W. W. Norton, 2006.

———. *Down to Earth: Nature's Role in American History.* New York: Oxford University Press, 2002.

Stine, Jeffrey K. "Placing Environmental History on Display." *Environmental History* 7, no. 4 (October 2002): 566–88.

Striffler, Steve. *In the Shadows of State and Capital: The United Fruit Company, Popular Struggle, and Agrarian Restructuring in Ecuador, 1900–1995.* Durham, N.C.: Duke University Press, 2002.

Sutter, Paul S. "Nature's Agents or Agents of Empire? Entomological Workers and Environmental Change during the Construction of the Panama Canal." *Isis* 98, no. 4 (December 2007): 724–54.

Taylor, Joseph E., III. "Boundary Terminology." *Environmental History* 13, no. 3 (July 2008): 454–82.

Taylor, Robert E. *Ahead of the Curve: Shaping New Solutions to Environmental Problems.* New York: Environmental Defense Fund, 1990.

Tomes, Nancy. *The Gospel of Germs: Men, Women, and the Microbe in American Life.* Cambridge, Mass.: Harvard University Press, 1998.

Train, Russell. *Politics, Pollution, and Pandas: An Environmental Memoir.* Washington, D.C.: Island Press, 2003.

Tren, Richard, and Liberty Institute. *When Politics Kills: Malaria and the DDT Story.* New Delhi: Liberty Institute, 2000.

Tren, Richard, and Roger Bate. "South Africa's War against Malaria: Lessons for the Developing World." Cato Policy Analysis, no. 513, 2004.

Tschinkel, Walter R. *The Fire Ants.* Cambridge, Mass.: Belknap Press of Harvard University Press, 2006.

Tsutsui, William M. "Landscapes in the Dark Valley: Toward an Environmental History of Wartime Japan," *Environmental History* 8, no. 2 (April 2003): 297–311.

———. "Looking Straight at THEM! Understanding the Big Bug Movies of the 1950s." *Environmental History* 12, no. 2 (2007): 237–53.

Turner, James Morton. "'The Specter of Environmentalism': Wilderness, Environmental Politics, and the Evolution of the New Right." *Journal of American History* 96, no. 1 (June 2009): 123–48.

Vagliano, Barbara. "Any Place but There: A Critique of U.S. Hazardous Export Policy." *Brooklyn Journal of International Law* 7 (1981): 329–65.

Van den Bosch, Robert. *The Pesticide Conspiracy*. Berkeley: University of California Press, 1989.

Vaudagna, Maurizio, ed. *The Place of Europe in American History: Twentieth-Century Perspectives*. Torino: OTTO editore, 2007.

Vogel, Sarah A. "Forum: From 'The Dose Makes the Poison' to 'The Timing Makes the Poison': Conceptualizing Risk in the Synthetic Age." *Environmental History* 13, no. 4 (October 2008): 667–73.

Vogt, William [Guillermo]. *El Hombre y la Tierra*. Mexico City: Biblioteca Enciclopédica Popular, Secretaría de Educación Pública, 1944.

————. *The Road to Survival: A Discussion of Food in Relation to the Problem of Growing Population*. New York: William Sloan Associates, 1948.

Waddell, Craig, ed. *And No Birds Sing: Rhetorical Analyses of Rachel Carson's Silent Spring*. Carbondale: Southern Illinois University Press, 2000.

Wagnleitner, Reinhold. *Coca-Colonization and the Cold War: The Cultural Mission of the United States in Austria After the Second World War*. Chapel Hill: University of North Carolina Press, 1994.

Wang, Jessica. *American Science in an Age of Anxiety: Scientists, Anticommunism, and the Cold War*. Chapel Hill: University of North Carolina Press, 1999.

Wang, Zuoyue. *In Sputnik's Shadow: The President's Science Advisory Committee and Cold War America*. New Brunswick, N.J.: Rutgers University Press, 2008.

Ward, Barbara, and René DuBois. *Only One Earth: The Care and Maintenance of a Small Planet*. New York: W. W. Norton, 1972.

Ward, Barbara, René DuBois, Thor Heyerdahl, Gunnar Myrdal, Carmen Miró, Lord Zuckerman, and Aurelio Peccei. *Who Speaks for Earth?* New York: W. W. Norton, 1973.

Weaver, Warren, ed. *The Scientists Speak*. New York: Boni and Gaer, 1947.

Webb, James L. A., Jr. *Humanity's Burden: A Global History of Malaria*. New York: Cambridge University Press, 2008.

Weindling, Paul Julian. *Epidemics and Genocide in Eastern Europe, 1890–1945*. Oxford: Oxford University Press, 2000.

————. *Nazi Medicine and the Nuremberg Trials: From Medical Warcrimes to Informed Consent*. Basingstoke, U.K.: Palgrave Macmillan, 2004.

Weintraub, Sidney. *A Marriage of Convenience: Relations Between Mexico and the United States*. New York: Oxford University Press, 1990.

Weir, David, and Mark Shapiro. *The Circle of Poison*. San Francisco: Institute for Food and Development Policy, 1981.

Wellford, Harrison, and the Center for Study of Responsive Law. *Sowing the Wind: A Report from Ralph Nader's Center for Study of Responsive Law on Food Safety and the Chemical Harvest*. New York: Bantam Books, 1973.

Westad, Odd Arne. *The Global Cold War: Third World Interventions and the Making of Our Times*. Cambridge: Cambridge University Press, 2006.

Wexler, Laura. *Tender Violence: Domestic Visions in an Age of U.S. Imperialism*. Chapel Hill: University of North Carolina Press, 2000.

Whelan, Elizabeth. *Toxic Terror: The Truth Behind the Cancer Scares.* 2nd ed. Buffalo, N.Y.: Prometheus Books, 1993.

Whitaker, Adelynne H. "Pesticide Use in Early Twentieth Century Animal Disease Control." *Agricultural History* 54, no. 1 (January 1980): 71–81.

Whitaker, John C. *Striking a Balance: Environment and Natural Resources Policy in the Nixon-Ford Years.* Washington, D.C.: American Enterprise Institute for Public Policy Research, 1976.

White, Donald W. "The 'American Century' in World History." *Journal of World History* 3, no. 1 (1992): 105–27.

White, Graham, and John Maze. *Henry A. Wallace: His Search for a New World Order.* Chapel Hill: University of North Carolina Press, 1995.

Whitten, Jamie L. *That We May Live.* Princeton, N.J.: D. Van Nostrand, 1966.

Whorton, James C. *Before Silent Spring: Pesticides and Public Health in Pre-DDT America.* Princeton, N.J.: Princeton University Press, 1975.

Williams, Robert Gregory. *Export Agriculture and the Crisis in Central America.* Chapel Hill: University of North Carolina Press, 1986.

Wilson, Clevo. "Environmental and Human Costs of Commercial Agricultural Production in South Asia." *Journal of Social Economics* 27, nos. 7–10 (2000): 816–46.

Worster, Donald. *Dust Bowl: The Southern Plains in the 1930s.* 25th anniversary ed. New York: Oxford University Press, 2004.

———. "Environmentalism Goes Global." *Diplomatic History* 32, no. 4 (2008): 639–41.

———. *Nature's Economy: A History of Ecological Ideas.* New York: Cambridge University Press, 1977.

———. "Transformations of the Earth: Toward an Agroecological Perspective in History." *Journal of American History* 76, no. 4 (March 1990): 1087–1106.

———. *The Wealth of Nature: Environmental History and the Ecological Imagination.* New York: Oxford University Press, 1994.

———. "World without Borders: The Internationalizing of Environmental History." *Environmental Review* 6, no. 2 (Autumn 1982): 8–13.

Wright, Angus Lindsay. *The Death of Ramón González: The Modern Agricultural Dilemma.* Austin: University of Texas Press, 1990.

———. "Rethinking the Circle of Poison: The Politics of Pesticide Poisoning among Mexican Farm Workers." *Latin American Perspectives* 13, no. 4 (Autumn 1986): 26–59.

Yoder, Andrew J. "Lessons from Stockholm: Evaluating the Global Convention on Persistent Organic Pollutants." *Indiana Journal of Global Legal Studies* 10, no. 2 (2003): 113–56.

Zachary, G. Pascal. *Endless Frontier: Vannevar Bush, Engineer of the American Century.* Cambridge, Mass.: MIT Press, 1999.

Zimmerman, O. T. *DDT: Killer of Killers.* Dover, N.H.: Industrial Research Service, 1946.

Zinsser, Hans. *Rats, Lice and History.* Boston: Little, Brown, 1936.

Index

Bush, Vannevar, 213 (n. 28); in World War II, 8, 12; vision articulated by, 8, 33–34, 47, 119
Butler, William, 154–55, 156–57, 158

Calvery, H. O., 31
Cancer, 13, 112, 144–45, 153
Candido, 50–52
Carbamates, 73, 139
"Carbaryl," 73
Carson, Rachel, 3, 97, 109–10, 114, 134–35. See also *Silent Spring*
Centers for Disease Control and Prevention, 13
CEQ (Council on Environmental Quality), 151, 163–64, 171, 176, 180
Chapman, Betty, 147, 218 (n. 40)
Chemical and Engineering News, 109, 120, 121
Chemical industry, 31, 32, 47, 72–73; European, 78, 166; and *Silent Spring*, 114–28 passim, 132, 134, 212 (n. 14); and debates over banning DDT, 137, 140, 142–43, 149, 153, 155, 158, 160. See also Exporting of pesticides; *specific companies*
Chemical poisoning. See Pesticide poisoning
Chemical Week, 109, 122–23, 126
Chile, 82, 129
China, 107, 163, 188
Chlordane, 125, 169
Chlorinated hydrocarbons, 72–73, 93–94, 97, 103, 139, 169. See also Aldrin; DDT; Dieldrin; Heptachlor
Cholera, 13
Christian Science Monitor, 31
CIBA-Geigy, 166
CIMMYT (International Maize and Wheat Improvement Center), 115
Cobb, William, 78
Coburn, Tom, 5
Cold War, 8, 36–37, 82, 134; DDT and, 8, 10–11, 47, 48, 55–56, 104, 108–9, 188; and science, 12, 118–20; and U.S. aid policies, 107, 124, 128
Colombia, 82, 129
Colwell, William, 67
Committee on Medical Research, 17
Commoner, Barry, 150–51
Competitive Enterprise Institute, 4–5
Congressional hearings, 5, 113–14, 132, 149–50, 175–76
Congress of Racial Equality, 5

Consolidated DDT hearings (1971–72), 151–59, 160, 163, 164
Consumer Product Safety Commission (CPSC), 175, 179
Cooper, Frederick, 130
Corn, 63, 174
Cottam, Clarence, 31
Council on Environmental Quality (CEQ), 151, 163–64, 171, 176, 180
Cox, Alvin J., 72, 76
Crichton, Michael, 5
Cuba, 23, 87
Cueto, Marcos, 65
Cyanamid Magazine, 124–25, 127–28

Daniel, Pete, 98
Dayton, Leon, 50, 51, 54, 58
D-D herbicide, 77
DDT (dichloro-diphenyl-trichloroethane), 1–11; efforts to ban, 4, 143–60 passim, 165–66; as exemplar of modern technologies, 6, 7, 8–9, 40, 61, 104, 184, 185; and American Century, 6, 8, 9–10, 102, 104, 138, 180, 188; emergence of, during World War II, 7, 14–34 passim, 40–46, 69; and Cold War, 8, 10–11, 47, 48, 55–56, 104, 108–9, 188; persistence of, 15, 104, 144, 163; origins of, 15–19, 20–21; use of, against typhus, 18, 29, 40–46; and mosquito eradication, 25, 48, 49, 52–53, 58, 66, 71, 90–91, 105, 156, 186; in Green Revolution, 63–64, 66, 69, 72, 73–74, 82–83; migration of, 102–3, 104, 163
"Decade of Development," 106, 107, 108, 137
Delaney Amendment (1958), 144
Dengue fever, 13, 14, 30
Deutschman, Paul, 56–57
Diamond, Edwin, 119
Diamond Shamrock, 165
Dichloro-diphenyl-trichloroethane. See DDT
Dieldrin, 72, 94, 112, 145, 169, 223 (n. 37); toxicity of, 72, 74, 112, 169; and Wisconsin DDT hearings, 138, 141; bans imposed on, 145, 169
Diphtheria, 13
Dow Chemical, 77, 166, 167, 206 (n. 69)
Downs, Wilbur, 65, 71–72, 91
Driessen, Paul, 5, 185–86
Droughts, 116, 128
Dubos, René, 165

González, Roberto, 81
Goodling, George A., 160
Good Neighbor Policy, 62
Gorgas, William, 87, 91, 105
Greece, 18, 93
Greenpeace, 138
Green Revolution, 61, 107–8, 129, 172; and technological modernity, 61, 108, 130; beginnings of, in Mexico, 61–82 passim, 115, 130; DDT in, 63–64, 66, 69, 72, 73–74, 82–83; Rockefeller Foundation and, 63–82 passim, 107, 115, 130, 172, 208 (n. 91); criticisms of, 70–71, 81, 107–8, 137, 187; Norman Borlaug and, 79, 115–16, 147–48, 155; and U.S. foreign aid, 107–8, 129, 176; in India, 128, 129, 130, 172; in Africa, 174
Guareschi, Giovannino, 50–52
Guatemala, 173
Gypsy moth, 16, 95–96, 103, 120

Hackett, Lewis, 47, 55
Haiti, 14, 175
Hall, Marcus, 49
Hanson, Frank, 73
Hardenburg, Walter Ernest, 88–90
Hardin, Clifford M., 143, 145, 220 (n. 65)
Harper's, 56–57
Harrar, J. George, 66–67, 68, 72, 77, 80–81
Helicopters, 1
Heptachlor, 97, 125, 169
Herter, Christian A., Jr., 165, 176
Hill, Archibald Vivian (A. V.), 98–99, 105
Horning, Donald, 129
Houseflies, 15, 93
Howard, Leland Ossian, 16
Hoy, Marjorie, 94
Hudson Spray Company, 45
Humphreys, Margaret, 23
Hunger, 115, 133–34
Hybrid varieties, 68, 79, 116, 174
Hynes, H. Patricia, 120

ICI, 166
Illinois, 103
Iltis, Hugh H., 141
Indonesia, 129, 175, 225 (n. 78)
Indoor residual spraying, 1, 42–43, 142, 184, 188–89
Insects, 14, 21, 31–32, 66, 86; and agricul-

ture, 68–70, 72, 73; scare stories about, 124–25
— pesticide resistance of, 2, 5, 58, 83, 93–95, 111–12, 148, 173; and goal of eradication, 93–95, 101, 102, 104–5, 143
See also Anopheles mosquitoes; Lice; Mosquito eradication
Institute of Inter-American Affairs (IIAA), 71–72
Integrated pest management, 112, 115, 157
International Corn and Wheat Research Institute (ICWRI), 115
International Maize and Wheat Improvement Center (CIMMYT), 115, 171
International Monetary Fund, 107
International Rice Research Institute (IRRI), 116, 172
Isle of Hope (film), 57
Italy, 37, 38–39, 54–55; malaria in, 21, 47–54; DDT use in, 28, 29, 34, 37, 40–46, 48–54, 55, 58–60; typhus in, 28–29, 39–46; in 1930s, 35–36, 37–38; Pontine marshes in, 36, 37–38, 39, 42, 55; typhus campaigns in, 40–46; DDT manufacture in, 180, 225 (n. 78). *See also* Sardinia

Jacobs, Jane, 85
Johnson, Lyndon B., 128, 129, 134; administration of, 56, 117, 128–29
Journal of Economic Entomology, 33
Jukes, Thomas H., 122–23, 149

Kennedy, John F., 106–7, 128; administration of, 56, 107, 117
Kerr, John, 48–49, 55
Kirk, Norman T., 30
Kissinger, Henry, 170–71
Knipling, Edward F., 16–17, 21, 33, 49, 96–98, 153–54; Orlando lab of, 16, 17, 18, 20–21, 22, 25, 28, 31, 33, 40; and screwworm campaign, 16, 96, 97, 154
Knowles, John H., 182
Kozol, Wendy, 39
Kronick, Paul, 121

Ladies Home Journal, 31
Larrick, George, 113
Latin America, 23, 62, 73, 94, 107, 129. *See also* specific countries

Lawns, 73, 110
Lead arsenate, 16, 69
Lear, Linda, 120
Lebanon Chemical, 165
Lefebvre, Henri, 84
Leonard, Jonathan, 120
Le Prince, Joseph, 87, 91, 105
Leptophos, 175, 179
Lice, 20, 27–28; DDT and, 20–21, 22, 29,
    40–46
Life, 35, 36, 51, 62, 80, 169; during World
    War II, 36, 39–46
Life expectancy, 13
Logan, John, 48–49, 53, 54, 58
Luce, Clare Booth, 36, 58
Luce, Henry, 36, 51, 58, 169; and "American
    Century" vision, 6, 34, 43–44, 59–60,
    80; and Italian fascism, 35–37, 38–39;
    publishing empire of, 36, 80, 169; in World
    War II, 38–39, 43–44
Lyle, Clay, 95–96
Lyng, Richard, 163

MacArthur, Douglas, 13, 29–30
Macy's, 54–55
Malaria, 2, 14, 19–20, 92, 150; transmission of,
    1, 19–20, 95, 174; methods of combatting,
    2, 17–18, 29–31, 37–38, 55, 91, 184 (*see
    also* Mosquito eradication); in Africa, 2,
    100–101, 174, 185; in United States, 13, 125;
    in World War II, 13–14, 17–20, 22, 25–27,
    29–31; in India, 56, 101–2, 150, 173
—rates of, 1, 2, 29, 30–31, 38, 173, 184, 224
    (n. 56); in Italy, 47, 48, 55
Malathion, 73, 102, 142, 156
Malkin, Michelle, 4–5
Malloy, Steve, 5
Mangelsdorf, Paul C., 64–65, 74, 78
Manzoni, Carletto, 51, 52
Marschesi, Marcello, 51
Marshall Plan, 50, 54, 55, 57
Mason, David, 188
May, Elaine Tyler, 121
McAlister, Melani, 10–11, 43
McCann, James, 174
McKelvey, John J., 67, 68
McLean, Louis A., 125–26, 142–43
McMillen, Wheeler, 132–33
Messersmith, George, 63

Metcalf, Robert, 95
Metcalfe, Elliot, 157–58
Mexican Agricultural Program (MAP), 65,
    66–69, 72–80, 82; ending of, 115
Mexico, 99, 130, 204 (n. 21), 206 (n. 69),
    209 (n. 25); DDT testing in, 22, 49, 65,
    209 (n. 25); and beginnings of Green
    Revolution, 61–82 passim, 102, 107, 115–16,
    130; pesticide poisoning in, 74–75
Michigan, 138
Migration: of mosquitoes, 89, 99; of DDT,
    102–3, 104, 163
Miller, Henry I., 185
Mirex, 97
Moeller, Susan, 46
Monsanto Chemical Corporation, 123–24, 166
Montrose Chemical Corporation, 155, 165–66,
    167, 175, 180, 188, 220 (n. 80); lawsuits
    against, 225–26 (n. 79). *See also* Rotrosen,
    Samuel
Moore, Robert, 14
Morell, Theodor, 18
Morton, Rogers C. B., 165
Mosca, Giovanni, 50–51
Mosquito eradication, 1, 85–91, 104–5, 186,
    219 (n. 56); DDT and, 1, 25, 48, 49, 52–53,
    58, 66, 71, 90–91, 105, 156, 186; and malaria
    rates, 1, 48, 173; and insect resistance, 2, 5,
    93–95, 101, 102, 104–5, 111–12, 143; critics
    of, 4–5, 48–49, 55, 59; and yellow fever,
    23, 24–25; Fred Soper and, 23, 24–25,
    27, 47, 48–49, 54, 90, 100; Rockefeller
    Foundation and, 25, 27, 48, 61, 65–66,
    71, 91; in Sardinia, 37, 47–54, 56–60, 85,
    90, 202 (n. 47); and Cold War ideology,
    55–56, 60–61, 85–86, 104, 121–22, 124; and
    American Century, 85, 86, 99, 101–2, 104;
    in Africa, 100–101. *See also* World Health
    Organization—Global Malaria Eradication
    Programme of
Mrak, Emil M., 144–45. *See also* Mrak
    Commission
Mrak Commission, 144–45, 147, 160
Müller, Paul, 7, 15, 18, 49
Multinational corporations, 116–17, 172. *See
    also specific corporations*
Mumford, Lewis, 84–85, 86, 105
Murphy, Priscilla Coit, 113
Mussolini, Benito, 35–36, 37–39

Mutual Security Agency, 57
MYL, 27–28

Nash, Linda, 112
National Agricultural Chemical Association
    (NACA), 126, 132, 139
National Association of Insecticide and
    Disinfectant Manufacturers, 32
National Audubon Society, 145. See also
    Audubon
National Cancer Institute, 145
National Environmental Policy Act (NEPA),
    151, 176
Neel, Paul A., 31
Nehru, Jawaharlal, 101–2
Nelson, Gaylord, 141
Nelson, Henry, 32
Neocid, 17–18, 20
Nepal, 175
New Deal, 13
Newsweek, 40, 109
New Yorker, 109, 120, 125, 126
New York Times, 3, 8, 28, 109, 110, 120, 167, 173;
    on DDT use, 29, 31, 148, 154, 158
Nicaragua, 146
Nicolle, Charles, 20
Nile River Valley, 26–27
Nivola, Constantino, 58–60
Nixon, Richard M., 151, 168; administration of,
    151, 162, 167, 168

Office of Malaria Control in War Areas, 13, 17,
    25–26
Office of Scientific Research and
    Development, 8, 32
Olin Mathieson, 165
Organochlorides. See Chlorinated
    hydrocarbons
Organophosphates, 73, 102, 139, 173, 187. See
    also Malathion
Osborne, Fairfield, 70

Paarlberg, Don, 115, 171
Packard, Randall, 130
Pakistan, 107, 128, 129, 150, 224 (n. 56)
Pampana, Emilio, 95, 99
Panama, 87
Pan American Health Organization, 49
Parathion, 73, 102

Paris green, 16, 26, 55, 69, 86–87, 88, 90;
    compared to DDT, 20
Parran, Thomas, 14
Pasteur Institute, 20, 27
Patterson, Gordon, 87–88
Peace Corps, 107
Pearce, Kimber Charles, 117
Perkovich, George, 102
Pesticide Action Network, 223 (n. 37)
Pesticide poisoning, 21, 74–75, 173, 187
Pesticide testing, 27–28, 72, 73–74, 113–14; of
    DDT, 22, 25–26, 72
Philippines, 115, 116, 172, 224 (n. 46)
Plasmodium falciparum, 19, 26
Plasmodium vivax, 19, 150
Poage, William R., 149, 161–63, 171
Poage-Wampler amendment, 170
Point Four Program, 56, 80–81, 101
Poisoning. See Pesticide poisoning
Polio, 13, 103
Pontine Marshes, 36, 37–38, 39, 42, 55
Population growth, 70, 98–99, 115, 116, 127–29,
    149–50
Prowazek, Stanislaus von, 20
Puleston, Dennis, 138
Putnam, Paul, 55
Pyrethrum, 15–16, 17, 20, 27, 69

Rarick, John, 149, 150
Raymond, Allen, 40
Reader's Digest, 40
Reed, Walter, 23, 25, 87, 91
Reimportation, 175–76
Resistance. See Insects — pesticide
    resistance of
Rhône-Poulenc, 166
Ribicoff, Abraham, 113, 114
Rice, 115, 116, 129
Ricketts, Howard Taylor, 20
Rickettsia prowazekii, 20
Risebrough, Robert, 141
Rockefeller, Laurence, 165
Rockefeller, Nelson, 71
Rockefeller Foundation, 9, 21–22, 55–56,
    71, 82, 91, 116, 117, 172, 204 (n. 21); and
    development of DDT, 22, 24–26, 31–32, 49,
    61; and malaria, 22, 26–27, 47, 48, 182, 209
    (n. 20); and typhus, 22, 27–28, 29, 41; and
    yellow fever, 23, 24; Fred Soper and, 23, 27,